BBC
BOOKS

COAST
THE WALKS

This book is published to accompany
the BBC television series *Coast*.

Published in 2008 by BBC Books,
an imprint of Ebury Publishing.
A Random House Group Company

10 9 8 7 6 5 4 3 2 1

Mapping extracts have been supplied by the Ordnance
Survey (© Crown copyright 2008); Ordnance Survey
Northern Ireland (Permit number 70230/© Crown
Copyright); Ordnance Survey Ireland, (Licence number
MP009307/© Ordnance Survey Ireland).

Map extracts for the rural walks are at a scale of 1:25,000
(4cm:1km), except the Causeway Coast which is at a scale
of 1:50,000 (2cm:1km). The town maps are not to scale.

While every effort has been made to provide the
most accurate route information, it is essential to take
the appropriate map with you on the walks. Buy your
maps online at www.ordnancesurvey.co.uk/mapshop
(for Great Britain), www.osni.gov.uk (for Northern Ireland)
and www.irishmaps.ie (for Ireland).

The Random House Group Limited Reg. No. 954009

Addresses for companies within the Random House Group
can be found at www.randomhouse.co.uk

A CIP catalogue record for this book is available from
the British Library.

ISBN 978 1 84 4607355 7

Commissioning editors:
Shirley Patton and Christopher Tinker
Production controller: David Brimble

Editing, design and layout by Butler & Tanner Printers Ltd,
Frome, Great Britain

Editor: Julian Flanders
Designer: Craig Stevens

Printed and bound by Firmengruppe APPL,
aprinta druck, Wemding, Germany

BBC Books would like to thank the Open University,
and in particular Diane Morris, for their help in
producing this book. To learn more about our coast
visit www.open2.net/coast

Thanks are also due to the Ordnance Survey
and to Robin Morley at the BBC for his help
in locating photographs.

To buy books by your favourite authors and register
for offers visit www.rbooks.co.uk

COAST
THE WALKS

OVER 50 WALKS INSPIRED BY THE BBC TELEVISION SERIES

Includes Ordnance Survey® mapping

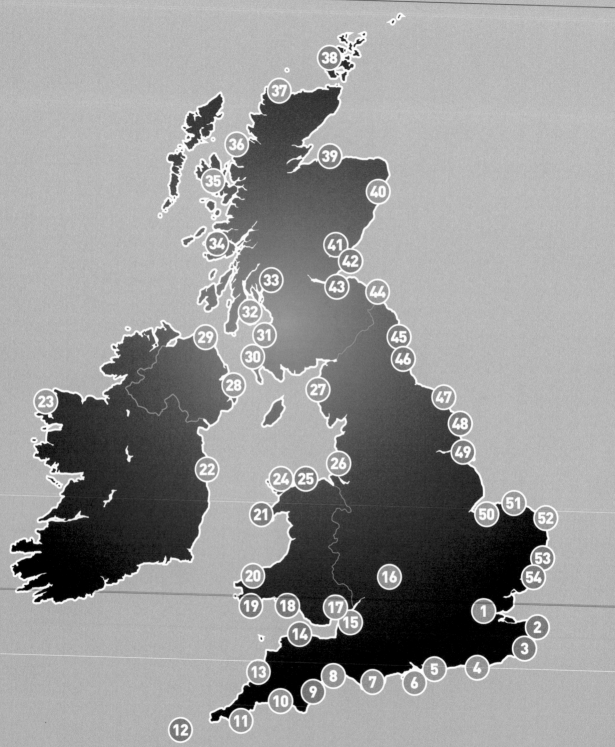

CONTENTS

THE WALKS

INTRODUCTION

Coast: The Walks is published to accompany BBC2's BAFTA award-winning television series *Coast*, and will inspire you to explore the best that the coastline of the British Isles has to offer. With more than 50 fantastic walks around Great Britain and Ireland, this practical guidebook features a variety of town and rural routes that are designed to suit both experienced walkers and those who fancy nothing more taxing than an afternoon stroll.

Whether you crave the rugged beauty of Antrim and Skye, the seaside fun of Brighton and Great Yarmouth, the majestic cliffs of the Cornish and Pembrokeshire peninsulas, or the history of our great coastal cities, such as Aberdeen, Belfast, Dublin, Hull, Liverpool and Plymouth, *Coast: The Walks* includes walks for everyone right the way around the coast. The walks are ordered geographically starting in the south-east and moving west along the south coast, then up through Wales and Ireland and western Scotland. There are some wild walks in the north of Scotland and some calmer ones moving south down the east coast through the Borders and Northumberland into White Cliffs country.

As well as offering invaluable advice that will enable you to plan your day out, and tried-and-tested instructions to guide you on your way, the book helpfully highlights points of interest along the route: historic sites and buildings, topographical features, and wildlife to watch out for as you go. Each walk is illustrated with colour photographs to help you understand what you will see when you get there.

A key at the start of each walk provides useful information about difficulty, accessibility, distance and timings. The level of difficulty is based on common sense. Easy walks are usually on tarmac or paved roadways and flat pathways; they are short and accessible to everyone. Medium walks tend to be longer or over rougher ground often with ascents and/or steep pathways. Hard walks are genuinely difficult, either because they are long or because they cover difficult and or dangerous terrain.

Read and take notice of the advice given in the panel at the end of each walk. which includes how to get to there, refreshment stops, short-cuts, plus useful telephone numbers and website details.

Dress appropriately, take a compass and a proper map of the whole area in case you need to find an alternative route to the one described, details of the relevant map are given on the map extract. Some of the walks traverse exposed sections of hillside and cliff where weather conditions can change dramatically in a very short space of time. Remember to take food and water, hats, sun cream or whatever is appropriate to your situation. A mobile phone is always helpful in case of emergencies. Some of the walks have warnings en route. Please take notice of them.

Equally at home in the car or on the bookshelf, *Coast: The Walks* will encourage you to get out there and make the most of our wonderfully diverse and endlessly fascinating 10,000 miles of coastline.

KEY

Difficulty

🚶🚶🚶 HARD

🚶🚶🚶 MEDIUM

🚶🚶🚶 EASY

Accessibility

🚶 ON FOOT

♿ WHEELCHAIR USERS

🚼 BUGGY/PUSHCHAIR USERS

Distance

👣 MILES

Time

🕐 HOURS:MINUTES

Approximate length of time the walk will take, excluding attractions that you may visit along the way

Rotherhithe
A Thames tour

EASY

ACCESS

3 MILES

2:00

WHALING

To cope with the increase in the whale trade Greenland Dock was fitted with boilers and tanks for extracting sperm oil from blubber brought in ships. The blubber was delivered in strips, which were then cut into small blocks and melted in iron pots. Whale oil was used to light homes and factories, to lubricate machinery and even for use in soap until the early 19th century. Whalebone was widely used in corsets, umbrellas and other products.

This walk along the famous River Thames will take you on an historic tour around the homes and workplaces of London's seafarers.

The walk begins close to Surrey Quays shopping centre under Redriff Road Bridge. As you walk under the bridge you enter **Greenland Dock**, part of Surrey Docks. The dock dates back to the 17th century and was originally known as Howland Great Wet Dock. Twice the current size and one of the largest docks in the world, it was used as a shelter for ships ready to unload their goods at the nearby legal quays.

Keep to the left as you walk around Brunswick Quay and Finland Street.

The dock was renamed Greenland Dock in 1763. Many ships sailed from here to Greenland, hunting whales for blubber and whalebone. You will notice on the left a pub named the Moby Dick, for obvious reasons. Whaling declined in the 19th century and gave way to timber and grain.

Timber, or deal, dominated Greenland Dock for the next 100 years. Imported from Scandinavia and the Baltic, the timber was unloaded by London's famous 'deal porters' – athletic men who carried and stacked timber.

Bombing during the Second World War devastated the docks, but they did revive until the timber trade ceased and the docks closed in 1970. A majority of the warehouses were demolished and rebuilt in the late 1980s. It is now mostly residential.

2 The walk continues towards the River Thames where you follow the Thames Path, taking in the views across to Westferry and Canary Wharf.

Turn left into Randall's Rents and right into Odessa Street before rejoining the Thames Path by the red crane. You will need to cut through Wyatt Close and Vaughan Street before turning right onto Rotherhithe Street – the longest street in London.

You will come to **Nelson House**, which was built in the 1740s on the site of a former shipyard. The roof has an octagonal cupola with stunning views of the river. This Grade II listed building is now used as offices and is not open to the public. Next to the house is Nelson Dock. This dry dock was used for building warships and clippers from the 17th century until it closed in 1968. The buildings you see today are the surviving sheds of ship repairers Mills and Knight.

As you continue along Rotherhithe Street you will pass the Blacksmith's Arms. If you are able to navigate steps you can turn right by the Canada Wharf building and walk along the Thames Path. Otherwise just continue along the street.

The next building of interest is the old fire station. Built in 1903 on the site of an older station, it was required to cope with any fires that might break out due to the large timber stocks located around the docks. The station was one of London's busiest and closed in 1965.

3 Just past the fire station is the **Pumphouse Educational Museum**, which also houses the Rotherhithe Heritage Museum. The pumphouse was built in 1929 to regulate the water level in the dock system, but today it is home to a collection of artefacts from Roman times to the Victorian era. The Rotherhithe Heritage Museum tells the story of Rotherhithe, which is one of the oldest villages in London.

After leaving the pumphouse continue along Rotherhithe Street. You can access the Thames Path between Globe Wharf and King and Queen Wharf and follow it until you exit by the Spice Island pub and the red swing bridge. There are no ramps where the walk exits by the pub, so if you cannot navigate stairs you should stay on Rotherhithe Street.

Globe Wharf is one of the finest warehouses left in the docks. It was built in 1883 as a grainstore and later used as a rice mill. King and Queen Wharf, next door, was built by French prisoners during the Napoleonic Wars in the 1790s.

4 As you walk down Rotherhithe Street on your left, in Railway Avenue, is the Brunel Museum on the site of the world's first underwater tunnel under a navigable river. Back in 1825 work began to build a

THE DEVIL'S TAVERN

While walking along the river here you can see across to Wapping where there are still many old warehouses and pubs. You should be able to see a white building – this is the Prospect of Whitby, one of London's oldest pubs, which dates back to 1520. In the 17th century it had a reputation as a meeting place for smugglers and villains, and became known as the 'Devil's Tavern'.

BRUNEL'S WAY

Before construction on the tunnel under the Thames could begin a huge shaft was built at ground level the same depth as the tunnel. Earth was removed from the shaft foundations allowing it to gradually sink down to its correct depth. Next to the shaft an engine house was built for the steam engines used to drain water from the tunnel. This unique shield tunnelling method was the first of its kind and the concept is still used today.

THE SHIPBUILDERS' CHURCH

St Mary's Rotherhithe dates back to the 12th century and was rebuilt by local shipbuilders in the Georgian era. The barrel roof was made to look like an upturned ship and the supporting pillars are complete tree trunks encased in plaster. The communion table and two chairs are made from the timbers of the 98-gun *Temeraire*, second in command at the Battle of Trafalgar.

tunnel under the Thames linking Rotherhithe and Wapping. Designed by Sir Marc Brunel, assisted by his son Isambard, the pedestrian tunnel took 18 years to complete. In 1869 it was converted into a railway tunnel to create the East London Railway. The tunnel became part of London Underground in 1948, it was then used by the East London Line, which is now closed for redevelopment.

5 Continue walking along Rotherhithe Street towards **St Mary's Church**. Three of the four owners of the *Mayflower* are buried in the churchyard, including Christopher Jones, captain when the ship sailed to America with the Pilgrim Fathers in 1620.

You will also find the tomb of Prince Lee Boo. In September 1782 three Rotherhithe men embarked for China aboard the East India Company's *Antelope*, under Captain Wilson. While passing a group of islands in the Pacific (later called the Palau Islands), they were shipwrecked during a storm.

They took refuge on the island of Ulong, striking up a relationship with the king, who gave them trees from which to build a new boat. The king watched as the ship was built using grindstones, anvils and other tools. He was so impressed with their skills that he asked the men to teach them to his son, Lee Boo. Three months later they set sail for China again, with Prince Lee Boo on board.

The ship, *Morse*, arrived back in England in July 1784 where Lee Boo was taken to the home of Captain Wilson in Paradise Row, Rotherhithe. He settled in London where he attended church and school, learned the language and dressed as an Englishman. In mid-December, only months after arriving, Lee Boo contracted smallpox. He died on 27 December and lies buried to the left of the church entrance.

6 Close by the church stands the **Mayflower pub**, so named because it was here that the *Mayflower* was fitted out for the transatlantic voyage that took the Pilgrim Fathers to the New World in 1620. The Pilgrims were a group of separatists who had broken

away from the Church of England. They decided to opt for a new life after hearing tales of earlier settlers in the Americas.

The Pilgrims sailed to Southampton on 5 August, finally departing from Plymouth on 6 September 1620 with 102 passengers aboard. Two months later, on 21 December, they landed in Plymouth Bay and became the first permanent European settlers in America.

7 Among the other interesting buildings here is **St Mary's Free School**. Originally founded in 1613 by Peter Hills and Robert Bell for the princely sum of £3 per annum, it is thought to be the oldest elementary school in London. Its purpose was to educate the sons of local seafarers.

Next to the school is the Watch House. Built in 1821 it was used by the local watchman or constable to keep an eye out for wrongdoers, particularly bodysnatchers. Watchmen wearing white overcoats and carrying lanterns were meant to be seen and heard, and they called the time and weather. Watchmen wearing blue were 'silent' and checked dark corners of the local area.

Bodysnatching was common in this area as surgeons at the local Guy's hospital required fresh corpses and body parts for medical research. Legally, only bodies of convicted criminals could be taken. In 1832 the Anatomy Act was passed, making it an offence to rob a grave. It was only legal to dissect the unclaimed bodies of people who had died in hospitals or poor houses.

8 Walk down Elephant Lane, past the Ship pub, where you can get excellent views across to Wapping. There is plenty of evidence of the trade that once thrived in this area. Tea, coffee, sugar, rum, spices, silks, furs and tobacco were just some of the cargoes brought into Wapping during the 18th and 19th centuries.

where justice was meted out to convicted pirates for over 400 years. The gallows were located on the riverbank and the tide would wash over the bodies three times before they were taken down. More notorious pirates, including Kidd, were left to hang in a gibbet, a type of metal cage, to deter other would-be criminals. Before you finish the walk, go through King's Stair gardens and back onto the Thames Path. Here you will find the ruins of a 14th century manor house and see the Angel pub. The pub is thought to be where J.M.W. Turner painted his famous painting *The Fighting Temeraire*.

Such cargoes attracted pirates. Across the river you can see the Captain Kidd pub, named after the famous pirate executed here in 1701. This was the site of Execution Dock,

Rotherhithe

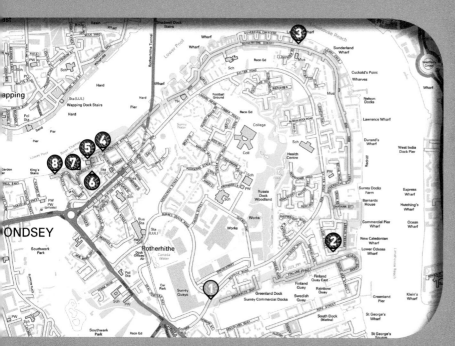

Ordnance Survey Explorer Map number 161
© Crown Copyright 2008

ADVICE

The walk is accessible to walkers, those in pushchairs and wheelchairs. There are a few of places with steps. These are mentioned in the text and an alternative route is suggested.

PARKING

You are advised not to use your car to get to the start of this walk. Parking is expensive and limited. You can get there by bus or take the Jubiliee Line to Canada Water.

START

The start point is Redriff Road Bridge, close to Surrey Quays or Canada Water tube. The end point is Elephant Lane, close to Rotherhithe or Canada Water tube.

CONTACT DETAILS

**Greenwich Tourist Information Centre,
Pepys House, 2 Cutty Sark Gardens,
Greenwich, London SE10 9LW**
t: **0870 608 2000**
e: **tic@greenwich.gov.uk**
w: **greenwich.gov.uk**

Dover
Gateway to the nation

MEDIUM

ACCESS

2 MILES

4:00

THE ROMAN FLEET

The soldiers and sailors of the *Classis Britannica* left Dover at the beginning of the third century to fight campaigns elsewhere, even though at this time the south coast was subjected to ever increasing raids by Saxon warriors from over the Channel in mainland Europe. The massive 10-foot width of the Saxon Shore Fort wall, which runs through the Roman Painted House, demonstrates how seriously the Romans regarded this threat.

Roman, Napoleonic and Second World War history are all waiting to be explored on this ramble round the strategically important coastal town of Dover.

❶ At the start of this walk you may want to linger on **Marine Parade**, especially if there is activity in the harbour. This could be a cruise liner arriving or departing, or a flotilla of yachts setting off for a race.

However, once you're ready to get underway, head westwards along Marine Parade towards the hoverport. If you look eastwards towards the sea, you can see the car ferries departing and arriving. On your right there is an enormous block of flats, called Gateway, which was built in 1958.

❷ Cross the road so you are alongside the Gateway gardens in front of the flats and you will soon reach the memorial to **Captain Webb**. Webb was the first person to swim across the English Channel and he did so on 24–25 August 1875. His swim lasted almost 22 hours, because he was pushed off course

by the tidal currents just off the coast of France. Unfortunately Captain Webb died barely eight years later, drowning when he tried to swim across the rapids below the Niagara Falls on the Canadian border in June 1883. Further along Marine Parade you'll come across another monument, a bronze statue of the Honourable Chas Rolls, who, on 2 June 1910, was the first person to fly across the channel to France and back without landing.

3 Continue your walk on Marine Parade until you see the white **Royal Cinque Ports Yacht Club Building**, then turn right into New Bridge and follow the signs to the town centre. You will pass the White Cliffs headquarters on your left.

4 Use the underpass to cross the very busy Townwall Street and proceed along King Street into **Market Square**, which has attractive raised flower beds and a fountain. You can relax here on one of the stone benches and look at Dover Castle in the distance or decide which of the cafes to choose for tea or coffee. If eating in the square, beware of the hungry seagulls, who will snatch your food or thump you on the head as they fly past at high speed.

The façade of the old market hall is Georgian and the building behind is the Dover Museum, where you can view the 3550-year-old Dover Bronze Age boat. It was discovered in 1992 by archaeologists from the Canterbury Archaeological Trust working with contractors building a new road between Dover and Folkestone. About half to two-thirds of the boat was recovered, 30 feet in total. It is made of oak timbers, held together by lengths of twisted yew, and the high quality of the workmanship shows that it could be used for crossing the English Channel. There is space for up to 18 paddlers and an estimated 3 tons of cargo. The boat highlights the crucial importance of the coast as a base for Bronze Age coastal and cross-channel trade and travel.

5 Having left the museum, turn left out of Market Square and walk down Cannon Street. Turn left into New Street where you will see the Roman Painted House on your left. Inside you can view the remains of a large Roman house, built 1800 years ago.

The house was located close to the abandoned Roman naval fort that was below the Western Heights, the headquarters of the Roman fleet, *Classis Britannica*. In 270 AD the Romans strengthened their coastal defences again by building the new Saxon Shore Fort, close to the old one. The Painted House was knocked down to make space for the fort's west wall, but demolition was incomplete, as the builders removed the roof and the upper storeys, but left the roofless ground floor rooms, which were filled with rubble and buried under the fort's ramparts. The rubble preserved the remains of the house until it was discovered in 1970 and the subsequent excavation lasted for 25 years. The inside surfaces of the remaining walls are still covered by painted plaster.

From the Roman Painted House, make your way to **Dover Castle** by retracing your steps along Cannon Street. When back in

CHALK-LOVING FLOWERS

The wild flowers growing in Dover Castle grounds vary with the months. For example, in August you can see the abundant pink flowers of wild marjoram and the bright yellow flower clusters of wild parsnip. There are also patches of the intense blue viper's bugloss and white ox-eye daisies. The chalky soil doesn't contain enough nutrients for grasses to grow, but the chalkland flowers flourish.

DICKENSIAN DETAIL

The Dickens' Corner tearoom on the corner of Market Square and Castle Street used to be the Igglesden and Graves Bakery and Restaurant. This is where Charles Dickens' character, David Copperfield, stopped to rest while searching for his aunt, Betsy Trotwood.

HELLFIRE CORNER

From the balcony outside the tunnel entrance at Dover Castle you can see Hellfire Corner, the most dangerous part of the route taken by the coastal convoys during the Second World War. The name came about because the Germans bombed and shelled the convoys relentlessly.

Market Square, turn left into Castle Street, walk past the Dickens Corner tearoom and head towards Dover castle, which you can see before you on top of the hill.

6 At the end of Castle Street, cross the road into Castle Hill Road and begin your uphill climb to the castle. Keep to the right-hand side of the road as there is a direct footpath from there that leads away from the road direct to the castle. This footpath is steep and includes a set of 86 stone steps. It's worth turning around on the platform halfway up the steps to view the Western Heights fortifications. They were built as defences against attack from France in Napoleonic times, from 1804 to 1808, and strengthened in the 1850s. However, long before that the Romans had occupied the site and built a **pharos or lighthouse** there.

7 At the top of the steps there is a kiosk on your left where you can buy your entrance ticket and, if required, book a guided tour of the **secret wartime tunnels**. Information about the castle and castle grounds is provided in the castle guidebook. From the ticket kiosk, walk across the drawbridge.

Your itinerary will vary according to your own taste, but it is worth starting in the Keep Yard and exploring the Keep first. This is a massive structure, four storeys high, built in the 12th century.

Your ticket also gives you entry to the secret wartime tunnels which were developed from tunnels excavated by the Royal Engineers in the late 18th and early 19th centuries, during the Napoleonic Wars. They were essentially an underground barracks used to house 2000 troops, ready to repel an invasion from France. On the outbreak of the Second World War, the tunnels were re-developed for various purposes. Operation Dynamo, the evacuation of the British Expeditionary Force (BEF) from Dunkirk in May 1940, was planned and coordinated in these tunnels, and brought back 338,000 men, the BEF and 139,000 French soldiers.

After the darkness of the tunnels it is worth exploring the castle grounds, which include grassy banks and meadows that are a riot of colour in summer when the chalk grassland wildflowers are blooming. Make your way towards the pharos close to the church of St Mary's in Castro. This lighthouse and the one at the Drop Redoubt guided

Roman ships and boats into the harbour. The pharos in the castle walls is 79 feet high, the tallest Roman structure in the UK. While viewing the sea from the meadow behind the pharos you may be able to see the French coast, 21 miles away.

8 Leave the castle by crossing the drawbridge and descend the steps. Walk a short distance and then turn left into Hubert Passage, a narrow lane that leads to the ruins of the old St James's Church. This church was built by the Saxons, but by 1860 it was too small for the growing numbers of worshippers. A new church was built in 1860 and the old church fell into disrepair. It was restored in 1869, but destroyed by German bombs in the Second World War, so there isn't much left to see, but the ruins provide an evocative memorial to the futility of war.

Turn left out of the ruined church and head towards Townwall Street, which you cross at a pedestrian crossing. You will see the back of the Gateway flats in front of you. Once over the road turn left and then right into a short road between the eastern side of the flats and the Premier Travel Inn, a large pale grey building. You will see Marine Parade in front of you and, again, you may want to stop here for a while. On a sunny day, it is pleasant to sit on the shingle beach and enjoy an ice cream.

Dover

ADVICE

Most of this walk is on town pavements and the buildings have at least some wheelchair access. Although most areas of Dover Castle are accessible to wheelchairs and buggies, pedestrian access to it is very steep, but you can drive up or take the no. 15 bus from Pencester Road Bus Station.

PARKING

There are pay-and-display bays, and toilets, in Marine Parade. There is a good choice of cafes in Market Square, not far into the walk.

If travelling by train, start the walk at the Roman Painted House. Turn left out of the station and follow Folkestone Road until you reach a roundabout.

Turn right into York Street and cross the road at the pedestrian crossing. Take the first left into New Street.

START

The walk starts and finishes on the seafront, in Marine Parade.

CONTACT DETAILS

**Dover Tourist Information Centre,
Old Town Gaol, Biggin Street,
Dover, Kent CT16 1DL
t: 01304 205108
f: 01304 245409
e: tic@doveruk.com
w: whitecliffscountry.org.uk**

Ordnance Survey Explorer Map number 138
© Crown Copyright 2008

DOVER

Saxon Shore Way
A Wealden walk

MEDIUM

ACCESS

3¾ MILES

4:00

As you walk along the Saxon Shore Way from West Hythe to Lympne and back again, across land that used to be covered by the ancient Wealden Lake, try to imagine the effects that geological change has had on this coastline over two millennia.

ROMAN FLEET PORT

The area around *Portus Lemanis* may have been a base for the Roman Fleet. The *Classis Britannica*, an altar stone, incorporated into the fort and discovered in 1850, is enscribed '[To the god] Neptune, Gaius Aufidius Pantera, prefect of the British Fleet [dedicates] this altar.' The fleet would have transported supplies for the army, particularly iron from the smelting areas in the Weald.

This walk provides an unusual experience as it explores a stretch of the Kent coast that has changed substantially within the period of recorded history. The route takes you through an area that was once part of the extensive Wealden Lake and the bed rocks are from the Cretaceous period, around 100 million years ago. In Roman times the sea covered almost all the land that now constitutes Romney Marsh and the car park where you start the walk was probably close to the old beach.

Leave the car park picnic area and walk west along the Roman shore line, now

the northern bank of the **Royal Military Canal**. You will notice that the bank is well above the water level of the canal and this was part of the original design. When Napoleon Bonaparte declared war on England in 1793 there was a fear that, if he attacked, Romney Marsh could not be flooded in time to act as a defence. So the 28-mile-long canal was constructed as a physical obstacle to invaders, with the excavated soil becoming a parapet on the landward side to protect the troops manning the defences.

② The parapet was originally stabilised by planting elms. On either side of the path you will find wild flowers that are typical of a clay soil. The Wealden clay underlies the canal and the marsh beyond. The yellow flowers of the common ragwort are prominent in the summer and autumn. The **common fleabane**, which is often found on clay soil as it prefers damp conditions, has flowers that look similar to ragwort, but with more petals. The flowers along the path attract insects. During the summer, look out for butterflies such as red admirals, peacocks and meadow browns.

③ After just under half a mile you reach a point where a canal cutting branches off towards the coast at Dymchurch. You can cross the canal, as a weir has been built at this point. If you look west along the canal, on one side you can see the **Roman shore** with the cliffs rising to the north, while on the other side is the flat land of Romney Marsh. This is a good place to try and picture the shore as it once was, with the sea being where the marsh land is now. A shingle spit used to extend from the cliffs at Fairlight up towards Hythe, so the sea was part of a sheltered lagoon. This is also a good place to watch birds. Kingfishers hunt all along the canal and there are also herons nesting in the woods. Swans and moorhens are common on the canal.

THE WEALDEN LAKE

In Cretaceous times this freshwater lake covered a large area of south-east England, northen France and Belgium. Rivers flowed into it, bringing silt and sand that formed the layers of rocks in the Weald today. The lake became connected to the sea and in this shallow marine environment the greensand was formed, the green colour coming from the mineral glauconite. Although the rocks formed at this time are found across north-western Europe, it is the Weald where they are most spectacularly exposed and that gives the formation its name – the Wealden.

ST STEPHEN'S LYMPNE

This beautiful church has a long history. The oldest part is the Norman tower, dating from 1100 AD. The rest of it appears to have been demolished and rebuilt by the 13th century. The west wall was entirely rebuilt in the 19th century and the church was restored after the Second World War following bomb damage.

LYMPNE CASTLE

This 13th-century castle was for many years the home of the Archdeacon of Canterbury. Archbishop Thomas à Beckett also lived here. By the end of the 19th century it was in poor condition, but was restored in 1905. The castle also had an important role as an observation post in the Second World War.

4 On leaving the weir rejoin the path along the parapet. As you walk along you should be able to see through the trees on your right, up on the skyline, the grey stone **Lympne Castle** and, lower down the hillside, the remains of a fort. Continue along the parapet until you reach a right turn onto a path that will take you up the old cliffs.

In the field to your right are the remains of *Portus Lemanis*, also known as Stutfall Castle, a Roman shore fort built in the 3rd century. Constructed of local sandstone, the fort provided protection for the coast and the entrance to the River Rother, which was navigable at that time. Of great interest is the fact that an altar stone to Neptune, encrusted with barnacles, was built into the fort. This must have come from an earlier building, possibly another shore fort, linked to the Roman fleet, the *Classis Britannica*. The fort had a relatively short life and by the 6th century land slips had already damaged it. None of the walls that are visible now are in their original positions and the shore edge that the fort once protected is now covered by clay that has slipped down from the cliffs above.

On your right are the formidable fences that mark the boundary of the late John Aspinall's zoo at Port Lympne. You may get glimpses of the animals in the distance.

5 The path gets steeper towards the top of the cliff, where the scarp formed by the greensand overlies the Atherfield clay, and joins the **Saxon Shore Way** path that runs along the cliff-top. Turn right and continue past the Millenium Wood on your left into the village of Lympne. Where you meet the road, you have a choice. Following the road as it curves away to the left will quickly bring you to the County Members pub for a break. Alternatively, turn right and follow the road along the edge of Lympne castle until you reach the lychgate of the church, St Stephen's Lympne, which dates from Norman times.

Enter the churchyard and walk down to the church and round to the back, where you can get a superb view out across Romney Marsh. As you sit on the seat in the graveyard looking out over the marsh, towards your left you can see the canal cutting snaking down to the sea at Dymchurch and, on a clear day, France is visible on the horizon. Straight ahead of you in the far distance are the massive buildings of the Dungeness nuclear power station and, just visible, there are two lighthouses to the left of them.

6 Leaving the churchyard, retrace your steps along the Saxon Shore Way path to the Millenium Wood and continue along the cliff-top, passing between enclosures that are part of the zoo, until you reach the Aldington Road.

7 When you get to the road, turn sharp left down the rough road and start your descent down the cliffs to return to the canal. There are great views out over **Romney Marsh** and on a good day Fairlight cliffs are just visible on the horizon. Most of the path is bordered on both sides by zoo paddocks, protected by high, wire fences. At the bottom of the cliff the path meets the canalside footpath. Turn left and follow the path back towards the starting point.

8 The use of the Royal Military Canal as a defence against invaders didn't end when the threat from Napoleon was over. In 1935 the canal was requisitioned by the War Office, to be fortified as a defence against invasion by the Germans. It was turned into a 28-mile anti-tank ditch and **pillboxes** were built on the parapet at strategic points, at least 15 in total. As you walk you will encounter a remakably well preserved polygonal one, now surrounded by trees. Continue along the path, past the weir and canal cutting, and return to the car park at West Hythe, completing your walk.

Saxon Shore Way

ADVICE

This walk is a mixture of easy and moderate walking. The route covers sections of the Saxon Shore Way and public footpaths and is open all year round. The section along the Royal Military Canal has a path and a bridle way, both of which are passable in dry weather with wheelchairs and buggies. The paths up the old cliff-side are steep, narrow and get slippery in wet weather. Dog walkers are asked not to allow their pets to foul the footpaths. There are picnic tables alongside the canal at the car park and there is a pub halfway round in Lympne.

PARKING

The car park at West Hythe is free, but spaces are limited. Some of the paths pass the Port Lympne wildlife park and are bordered by secure fences.

START

The car park at West Hythe is your starting and finishing point.

CONTACT DETAILS

Ashford Visitor Information Centre,
18 The Churchyard, Ashford, Kent TN23 1QG
t: 01233 629165
e: tourism@ashford.gov.uk
w: ashford.gov.uk

Ordnance Survey Explorer Map number 138
© Crown Copyright 2008

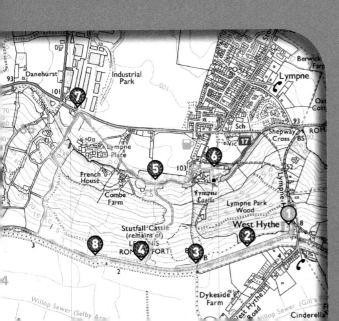

Brighton
A day-tripper's paradise

EASY

ACCESS

6 MILES

3:30

①

ST PAUL'S CHURCH

Built in 1848 as a mission church for the fisher folk of Brighton, St Paul's was a leading church in the Tractarian (or Oxford) Movement and that 'high church' legacy is visible today in the traditional Catholic appearance of the building's interior.

As much about soaking up the atmosphere as seeing the sights, this circular walk enables you to experience the traditional British seaside and explore the history of a famous south-coast resort.

It was awarded city status in 2000, but Brighton's origins date back to before the Domesday Book. The ancient settlement of Brighthelmston was a fishing community and the central green area, the Old Steine, was once the place where fishing nets were hung out to dry. However, during the 18th century Brighton started to become the popular, cosmopolitan resort it is today when the Prince Regent (later King George IV) chose it as his retreat. Brighton's most famous building, the Royal Pavilion, was his legacy. In the 19th century, Brighton's reputation as a destination for day-trippers grew, particularly when the direct rail link from London opened in 1841, and, fittingly, it is at the railway station that this walk begins.

① Leave the station's main entrance and walk down the right-hand side of Queen's Road, which then becomes West Street. The **sea front** soon comes into view at the end of the street. On you right you pass the church of St Paul's and immediately after the church is a covered passage, reminiscent of a cloister, which is the way in. To have a look at the church, which is open to visitors most mornings and some afternoons, walk up the passage and go through the door on the right.

2 Back on West Street, continue down to the sea front, which will obviously be much busier in summer than it is in winter, and cross the main road onto the promenade. Take the slope to the right down onto the lower promenade and wander among the

little shops set into the promenade wall, fronting the shingle beach. Head towards the pier until you reach a number of old fishing boats on display. Take a break from your walk and go into the **Brighton Fisheries Museum**, which traces the long history of fishing in Brighton. Just inside the door of the museum is a typical Sussex fishing boat, a clinker-built punt, which is about 30 feet long.

3 From the museum head back onto the upper promenade, towards the pier. Standing at the entrance to the new **Palace Pier** in 1899 you would have seen the remains of the Chain Pier to your left, destroyed in a gale in 1896, and the West Pier to your right. There is no trace now of the Chain Pier and what is left of the West Pier is a sorry blackened skeleton offshore.

The remaining pier is packed with buildings now, but it is still possible to take a gentle stroll along its outside edge, to the amusements at the end, round and back. There are shelters along the way where you can sit and contemplate the view, while trying to imagine what the atmosphere on the pier might have been like in late Victorian times, before the arrival of the modern amusements and the ever-present piped music.

4 Turn right as you leave the pier and walk down to Madeira Drive, which runs below the main promenade on Marine Parade. Pass the Sea Life Centre on your left and then on your right the Aquarium Station for **Volk's Electric Railway**. You will return to both of these later, but for now continue along the promenade and as you walk look at the plants growing alongside the railway.

As you approach the halfway station at Paston Place (formerly Peter Pan's Playground), you will see a sign on the railway fence telling you about the native plants that are characteristic of the shingle here. Most of the shingle behind the area covered by the tide has been built on, but this place has been preserved and designated a site of special scientific interest.

VOLK'S ELECTRIC RAILWAY

The son of a German clockmaker, inventor extraordinaire Magnus Volk was born in Brighton in 1851. He was one of the first people in Brighton to equip his house with electric light, which led to the contract for providing the Royal Pavilion with electric incandescent lighting. It wasn't quite the first electric railway in the world, but his most ambitious project, the Volk Electric Railway, was the first in Britain and, with the exception of the First World War, it has been running since 1883, making it the oldest remaining operating electric railway on the planet.

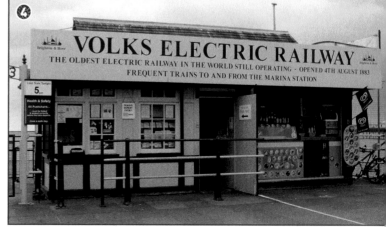

SHINGLE PLANTLIFE

You will find an amazing variety of plants growing amongst the shingle. This includes: black nightshade, bristly ox tongue, ivy-leaved toadflax, orache, red valerian, sea kale, sea beet, silver ragwort, woody nightshade and yellow-horned poppy.

A TALE OF THREE PIERS

Brighton's first pier was the Chain Pier, which opened in 1823. It was followed by the West Pier in 1866 and the Palace Pier in 1899. In addition, there was Magnus Volk's Brighton-to-Rottingdean electric railway, which was a moving pier that ran on underwater rails from the Palace Pier, between 1896 and 1901. Only the Palace Pier now remains.

⑤ Cross the parking area and walk under the **overhanging promenade** to look at the trees. These have been planted so that the trunks grow up the wall and squeeze between the edge of the next promenade level and the cliff side. All the trees' leaves are on branches above the next level. The original offices of Volk's Electric Railway, which was founded by one Magnus Volk and opened in 1883, are built into the side of the cliff opposite the station. They are still in use.

⑥ Just beyond the station, cross the railway at the **level crossing**. This is a good point to get a closer view of the trains and, if you want to, take photographs of them. Two two-car trains normally operate on the line, one in each direction, and their cross-over point is in the station. Continue your walk between the railway and the bank of shingle protecting the nudist beach, still keeping a look out for the plants.

⑦ When you reach Black Rock Station, where the railway line ends, cross the car park and follow the cycle route up through a tunnel that emerges on the cliff-top. Turn

right and follow the cliff-edge path to get a good view of the Brighton Marina.

You now have a choice. If you are feeling fit, cross the road and walk back along Marine Parade to the Sea Life Centre, admiring the sea front houses of Kemp Town and the **Royal Crescent** as you go.

Thomas Kemp MP started the construction of Kemp Town in 1823 as a Regency-inspired area of crescents and terraces. These houses became the social centre of Brighton up until the Second World War. Look out for the blue plaque at 160 Marine Parade that commemorates the comedian Max Miller, who once lived in the house. Just beyond is the Royal Crescent. This is a true Regency creation, started in 1799 by a West Indian merchant. The actor Laurence Olivier lived at number 4 and there is a plaque on the wall noting this fact.

Alternatively, return to Black Rock Station and wait for the next Volk's Electric Railway train back to the Aquarium Station.

⑧ If you have time, the Sea Life Centre itself, a modern acquarium housed in Victorian splendour, is well worth a visit. Continue the walk along the sea front, past the end of the pier, to the first pedestrian crossing, signposted to the Royal Pavilion. Cross into East Street and continue through the pedestrian area to the archway entrance to the grounds of the **Royal Pavilion**. To get the best view walk around this stunning building to the entrance on the other side. The bizarre oriental palace started out as a farmhouse, which, in 1787, was highly modified to produce the Marine Pavilion. John Nash later built an iron framework over the original pavilion and added all the orient-style excrescences that give the Royal Pavilion its unique character. Built for George IV and used by William IV and Victoria, the interior became as extravagantly royal as the exterior.

Leave the Pavilion grounds through the archway and at the end of the street turn right into North Street. Follow this until it meets Queen's Road and turn right to head back up to the station and the end of your walk.

Brighton

ADVICE

Walking around Brighton is easy and mostly on the level. The entire walk is accessible to wheelchairs and buggies, and there is no shortage of refreshment stops.

PARKING

Parking is difficult, but there is a multi-storey car park at the Brighton Centre on the sea front and limited parking immediately behind the beach between the Aquarium and the Marina.

START

The walk begins – and ends – at Brighton railway station.

CONTACT DETAILS

Brighton Visitor Information Centre, Royal Pavilion Shops, Pavilion Buildings, Brighton, East Sussex BN1 1EE
t: 0906 7112255
w: visitbrighton.com

Ordnance Survey Explorer Map number 122
© Crown Copyright 2008

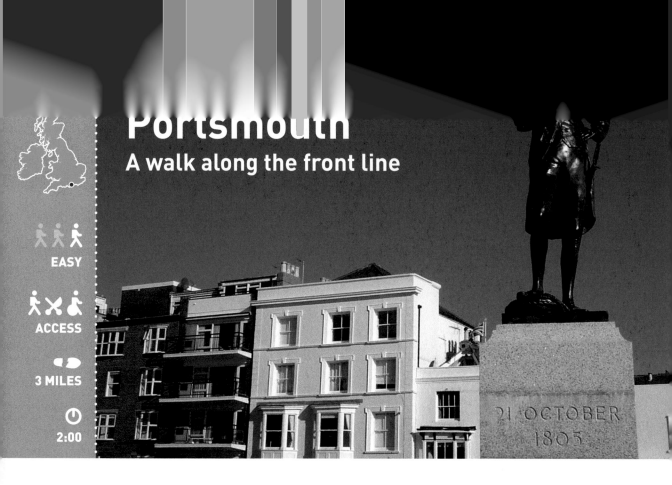

Portsmouth
A walk along the front line

EASY

ACCESS

3 MILES

2:00

21 OCTOBER 1805

THE RAISING OF THE *MARY ROSE*

This famous warship remained at the bottom of the Solent until she was spotted during an underwater survey in 1968. Finally raised in 1982, the *Mary Rose* is now preserved in her own museum at Portsmouth's Historic Dockyard along with many other items uncovered during excavation, including navigational and medical equipment, carpentry tools, guns, cooking and eating utensils, lanterns, backgammon boards and playing dice.

The nautical town of Portsmouth would have been a very different place to stroll a few hundred years ago – it would have been heaving with rowdy sailors and soldiers, and 18th-century superstar Horatio Nelson would have been a familiar face.

❶

The walk starts at Clarence Esplanade on the seafront. Portsmouth has always been on the front line of invasion, and on 19 July 1545 England was threatened by a huge fleet of over 200 French ships heading up the Solent. Henry VIII's high-tech warship, the *Mary Rose*, set sail from the harbour to meet the rest of the English fleet of around 80 vessels and join the battle.

Thousands of people had gathered here to see the enormous ship off. But as she hoisted her sails, she heeled over to port in the wind. Already heavy with over 400 crew on board

and the weight of the guns on her decks, she heeled further until her gunports were below the sea. Her hold quickly filled with water and she sank.

Watching helplessly from the seafront the king, his courtiers, the army and the crowd of well-wishers could do nothing as the crew (few of whom could swim) threw themselves into the sea – only 35 survived.

2 Continue walking towards Clarence Pier. In the spring of 1944 Portsmouth was at the centre of the preparations for D-Day. The **stretch of water in front of the pier** was a mass of ships all destined to take part in the greatest seaborne invasion in history.

The battle was directed from an underground operations room where Elsie Horton was stationed. This is a part of her account of the preparations. 'In 1944 I was a Wren on the staff of the Commander-in-Chief at Fort Southwick on Portsdown Hill. At about 22.00 hours on the night of Saturday 3 June, a signal gave news of a convoy sailing from the West Country, its destination 'Far Shore'– a name translated as the coast of Normandy. This was followed throughout the night by similar ship movements; it was obvious that the 'Second Front', so long awaited, was at last taking place.

'On the Monday [5 June] I was on late watch. As all the signals relating to D-Day were secret, only a few of us were aware of impending events. We were relieved by the night watch and returned to our quarters in Fort Wallington. I had a sleepless night, listening to the roar of planes overhead, and thinking of all the men on their way to battle, and what they might face.'

3 Continue along the seafront, walking away from the pier towards the wooden bridge. You're standing on Spur Redoubt one of the outer fortifications of Portsmouth. The tunnel or 'Sally Port' you can see through

the ramparts is the route Lord Admiral Horatio Nelson took on his way to board HMS *Victory* before the Battle of Trafalgar on 14 September 1805.

The statue of Lord Admiral Horatio Nelson wearing the uniform he wore when leaving Portsmouth that day stands on the spot where he took his last steps on dry land. Nelson took an unusual route to the beach that day to avoid the huge crowds that had gathered in the High Street to see him off. Leaving via the back door from the George Hotel in the High Street, where he'd spent the night, **Nelson headed along Pembroke Road** and cut across the green, past the Garrison Church through to the ramparts.

4 Continue until you reach a large paved area with lots of benches. Her Majesty's Naval Base here at Portsmouth is the home of the Royal Navy and houses two-thirds of the navy's surface fleet. This area is where navy VIPs salute fleets of ships as they sail off to war and where families and well-wishers greet the crews when they return home. It's known as the **Saluting Platform**.

DEATH OF A HERO

A month after leaving Portsmouth, on 21 October 1805, at the height of the Battle of Trafalgar where he led the British fleet to victory over the French and the Spanish, Admiral of the Fleet Lord Nelson was struck by a French sniper's bullet on board the frigate *Redoubtable* and later died. It was appropriate that the last dry land he stood on was that of the town most associated with England's naval prowess.

CALM AFTER THE STORM

Elsie Horton returned to work the following morning, Tuesday, 6 June, having heard on BBC radio that the troops had landed in France. 'It was strange', she said, 'to look down over Portsmouth and see the harbour, which had been so full, now empty of all shipping. Only the *Victory* was still there – was it a hopeful signal?'

(6)

FANCY A CUPPA?

In the 17th century the Garrison Church was where Charles II married Catherine of Braganza, a Portuguese princess. She had a very rough sea voyage over from Portugal and on arrival asked the locals if she could have something to drink. Being Portsmouth they gave her a jug of ale, which she didn't like. She sent out to her ship for some supplies to keep her going while she waited for her fiancé to arrive and they came back with some tea. It's possible that this was the first cup of tea drunk in Britain.

(5)

This area was packed on 21 July 1982 when HMS *Hermes*, the flagship of the British taskforce to the Falklands, arrived back after ten months away in the South Atlantic. After 108 days at sea, the aircraft carrier, named after the messenger of the Greek gods, sailed back into Portsmouth after her 8000-mile mission. Dirty and streaked with rust, she arrived with one side decorated with a scoreboard showing the 46 enemy aircraft shot down by her Sea Harrier fighters.

(5) Make your way down the ramp and walk along the narrow cobbled street into Broad Street. From here you can see the remains of an enormous gate, which 250 years ago marked the dividing line between the garrison and the rest of town.

King James's Gate sat within a huge stone arch and was manned by soldiers to keep the sailors apart from the 'respectable' townspeople. Originally there was also a small moat with a drawbridge – the soldiers used to close the gates and raise the drawbridge in the evenings, closing off the part of the town known as Spice Island.

Spice Island is where Portsmouth's shipping history began. With its perfect natural harbour, it quickly became a flourishing port and was soon full of the sailors' favourite facilities – liquor houses, taverns, pawnbrokers and brothels. It was open 24 hours a day and was a favourite haunt of pickpockets and press gangs.

(6) Follow the road along to the square tower and then turn left up the High Street towards the **cathedral**. The impressive stone building in front of you was built when Portsmouth became a city in 1926. Before that there was a small parish church here and a very popular pub, which fuelled one of the most dramatic nights in Portsmouth's history.

Founded in 1180 by a wealthy Norman merchant, Jean de Gisors, Portsmouth's natural harbour saw it soon develop into a garrison town. In the 1680s the town was little more than a fortified military post occupied by thousands of soldiers and sailors.

In 1685 a troop of Irish soldiers were in Portsmouth to support James II, the last Catholic king of England. Unfortunately for them, he had fled to France and they were not being paid. Having spent a day in the pub they rolled into the nearby church and started firing shots before rioting down the High Street.

(7) Continue walking to the roofless **Garrison Church**. Known as the *Domus Dei* (God's House), this building started life in 1212 as a hospice and shelter for pilgrims from overseas who were heading for the holy shrines at Canterbury, Chichester and Winchester. On the night of 10 January 1941 a firebomb raid on Portsmouth gutted the nave of the church leaving it roofless. It is now a memorial to the Blitz.

The area got a bad name in January 1449 when the Bishop of Chichester, Adam Moleyns, was murdered near the church's entrance by a mutinous group of seamen after an argument over their wages. For this chilling crime, Portsmouth was excommunicated – which meant that religious services couldn't be held in the town for 50 years.

(8) Continue the walk back across the Common into Pier Road. In the late 1800s

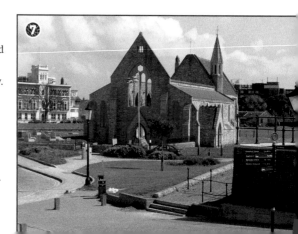
(7)

stretching your legs in the afternoon with a stroll along the seafront here before taking tea in one of the many cafes around the Common was a popular pastime for Portsmouth's high society crowd. The opening of the Southsea Baths meant that bathing in style and comfort was added to the list of leisure pursuits.

Unfortunately some of the town's working men chose the same place to indulge in a spot of nude bathing. To put a stop to this inappropriate behaviour the Pier Company, who owned the Baths and its tearoom, put up a barrier stopping free access to the Common and the beach.

Local councillor Barney Miller was furious and gathered a crowd to rip the barrier down. Over 5000 people turned up with axes, saws and hammers and started demolishing the fence in front of the tearoom's windows, where the tea drinkers had taken shelter.

The struggle lasted for four days until the army were eventually called in to restore order, but the people had had their day – the Battle of Southsea was won and the barrier was gone.

It's a short walk from here to the car park behind Clarence Esplanade, or simply turn left along the seafront to return to the starting point.

Portsmouth

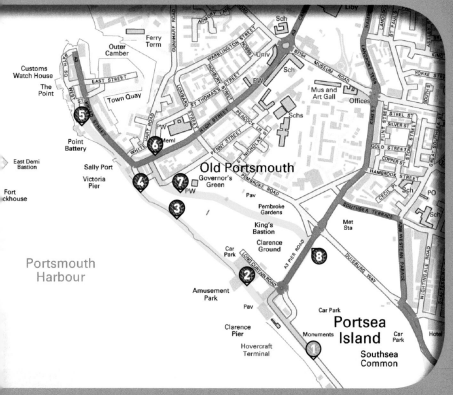

Ordnance Survey Explorer Map number 119
© Crown Copyright 2008

ADVICE

The whole walk is on pavement and tarmac and is fully accessible for wheelchair users and pushchairs/buggies.

PARKING

There is a pay-and-display car park on Clarence Esplanade in Southsea, on the corner of Southsea Common, where Pier Road joins Clarence Esplanade.

START

Follow the brown signs to Portsmouth Cathedral off the High Street. The walk starts on Clarence Esplanade.

CONTACT DETAILS

Southsea Visitor Information Service, Clarence Esplanade, Southsea, Hampshire PO5 3PB
t: 02392 826722
f: 02392 827519
e: vis@portsmouthcc.gov.uk
w: visitportsmouth.co.uk

Isle of Wight
The Needles Old Battery

EASY

ACCESS

4½ MILES

4:30

THE NEEDLES

These comprise three 100-foot pinnacles of chalk standing in the sea and are all that remains of a ridge that joined the Isle of Wight to the mainland. A taller and narrower fourth pinnacle, known as Lot's Wife, collapsed in 1794 and it was from this that the Needles get their name. The remaining rocks are not needle-like at all. The lighthouse on the tip of the rocks, designed by Scottish civil engineer James Walker, was built in 1859 to replace one on the cliff-top.

This walk provides spectacular views of a chalk coast and, from the top of the cliffs, superb vistas of the Isle of Wight, the Solent and the mainland.

1

The Isle of Wight only became an island about 7000 years ago after the last glacial period. The area over which you will walk is chalk, formed around 85 million years ago in the Cretaceous period. It is composed of the remains of planktonic organisms accumulated over millions of years when the area was below the sea. Later, when the area was covered by shallow coastal waters, the sands and clays that form the cliffs in Alum Bay were laid down. At the Alum Bay bus stop there is a National Trust signpost directing you up a path that runs through the car park and then on to a gated road that leads

towards the Needles. The only traffic on this road is island tour buses. At the top of the first slope the road bends sharply to the right and on the left you will see a sign for the path to Tennyson Down. You will be returning down this path at the end of the walk.

Meanwhile, follow the road, or the chalk path that runs alongside it, towards the Needles. On your right the land slopes down towards the cliff-edge, providing a good selection of wild flowers typical of the chalk landscape. As you get further along the path you will start to get good views of **Alum Bay** and the brightly coloured streaks of the sands that form the cliffs there. The red and brown sands are produced by the rusting of iron-rich minerals. Pure quartz produces the white and glauconite the green. Blue, purple, black and yellow can also be found.

2 Continue along the road until you reach the entrance to the Needles Old Battery (now owned by the National Trust). The battery was completed in 1863 for coastal defence against a threat of invasion by France and remained in use for various purposes until the Auxillary Coastguards left in 1986.

A ditch protects the battery on the landward side and you pass over the drawbridge to enter. You can see the original winding gear by the entrance. There are a number of rooms that you can visit, but the centrepiece of the battery is the pair of original **muzzle-loading guns**, of which six were installed between 1873 and 1893. They are mounted on replica carriages. The guns were obsolete by 1903 and were simply thrown over the cliff. The two in place now were recovered from the sea in 1983. In the middle of the parade ground, behind the guns, is a spiral staircase going down to a

tunnel through the cliff to the searchlight station overlooking the Needles.

3 As you leave the gateway of the Needles Old Battery look down to your left and you will see buildings in the ditch, below the road level. These housed the Robey steam engine that provided power for the searchlights. On your right, steps and then a path lead up the cliff towards the new coastguard lookout. Follow the path to the left of the lookout and you emerge into the **New Needles Battery**, built between 1893 and 1895 after it became apparent that the firing of the heavy guns was causing the cliffs to erode rapidly. The New Battery was in service up until 1954, when the site was leased by Saunders-Roe for the establishment of a secret rocket test site for the UK space and defence programmes.

4 Turn to your left and, at the edge of the gun position, go down the slope into the old command centre for the guns, which later became the **command centre for the rocket tests** (now also owned by the National Trust). These damp rooms housed the sophisticated electronics needed for the rocket tests and the heating was on permanently to keep things as dry as possible. There is a small video theatre where you can watch filmed interviews with the rocket scientists and a tea room for refreshments afterwards.

THE NEEDLES SEARCHLIGHT

The searchlight is sited in the cliffs above the Needles and when in use could illuminate ships passing through the Needles Passage. There are excellent views from the searchlight platform.

ROCKET TESTS

Saunders-Roe leased the New Battery site for testing the engines of the Black Knight and Black Arrow rockets, which they manufactured for the UK defence and space projects. There are two concrete bases for the test pads left. The rockets were clamped upright on these and the engines were test-fired. Three thousand gallons of water per minute were used to cool the exhaust gases that emerged horizontally from the test pads.

⑤ As you emerge from under the ground follow the road to the right towards the coastguard lookout, which is built on top of one of the gun positions. Follow the road round behind it and you will reach the two **rocket test sites**. These were used for static tests of the rockets, before they were shipped to Woomera in Australia for launching. Looking at the massive concrete structures it is hard to imagine that this test site was highly secret and yet was in sight of the Needles, a great tourist attraction. Keep to the paths here, as the cliff needs to be protected from erosion by visitors' feet.

⑥ Walk back along the road behind the gun positions until you reach a path that runs up alongside the old Coastguard cottages. Walk up the side of the cottages and follow the grass path across to a fence and stile. On the other side of the fence you have a choice of two grass paths: one straight on following the cliff-edge and one to the right following the crest of the chalk. Both bring you to the base of the hill upon which Tennyson's monument stands – visible in the far distance and seemingly a long way away. The distance is deceptive and less than an hour will bring you to the monument. From the crest of the chalk you can see across the Solent to the mainland and the views on a clear day are stunning, so don't forget to stop and look around at intervals. At the base of the hill leading up to the monument is what remains of the **old beacon**, which was removed when the monument was built.

ALFRED LORD TENNYSON

Tennyson lived at Farringford, just below Tennyson Down, for nearly 40 years, until his death in 1892, and he wrote much of his most important work here. He had many famous visitors to the house, some of whom were photographed by his neighbour, and renowned portrait photographer, Julia Margaret Cameron. The house is now a hotel.

⑦ Climb the hill towards the monument. On your right there are areas of **bell heather** (ling) that give the hillside an attractive purple colour in summer. This isan unusual chalk heathland and the soil must be slightly acid for the heather to flourish. As you approach the monument, you will notice that the cliff edge is slumping and the monument may eventually have to be moved. It is a 38-foot-high Ionan cross in Cornish granite, dedicated in 1897 by the Archbishop of Canterbury. Tennyson's house is in the valley below, towards Freshwater, to the east of the monument. Take a moment to smell the fresh, clean air. The poet would walk the down almost every day, saying that the air was worth 'sixpence a pint'.

⑧ Walk back down the hill to the stile by the old beacon and take the right-hand ork of the path down into the valley. Follow it, passing a farmhouse that offers teas, until you reach the road to the Needles, where you turn right and head down to the Alum Bay bus stop. If you have time before the bus, follow the path through the shops and stalls to the cliff-edge where you can see the **monument to Marconi's early work** with wireless at this place, which is explained on the four panels.

Isle of Wight

ADVICE

The route covers sections of the Isle of Wight Costal Path and is open all year round. The first part of the route is easy and accessible for pushchairs, but there is a steeper climb up to the cliff-top, before the easy walk across the downs. There is continuing erosion of the cliffs and you will see notices asking you to keep to marked paths in certain areas. Don't be tempted to look over cliff-edges as not only is it a long way down, but the edges are also liable to crumble. The large Alum Bay visitor area (Needles Park) has plenty of facilities, the Needles Old Battery has a cafe and there is a farm offering teas on the last stage of the walk. The walk offers a mixture of wildlife, history, technology and scenic views and is obviously best when the weather is good with no mist.

PARKING

There is a pay-and-display car park at Alum Bay, but no vehicle access to the Needles area beyond that. There is also ready access to public transport: trains from Brockenhurst connect directly with the ferries at Lymington that cross to Yarmouth. Buses from Yarmouth bus station run to Alum Bay.

START

The walk starts and ends at the car park at Alum Bay.

CONTACT DETAILS

Isle of Wight Tourist Guide Ltd,
17 Albert Road, Cowes,
Isle of Wight PO31 8JU
t: **01983 292746**
e: **info@isleofwighttouristguide.com**
w: **isleofwighttouristguide.com**

Ordnance Survey Explorer Map number OL29
© Crown Copyright 2008

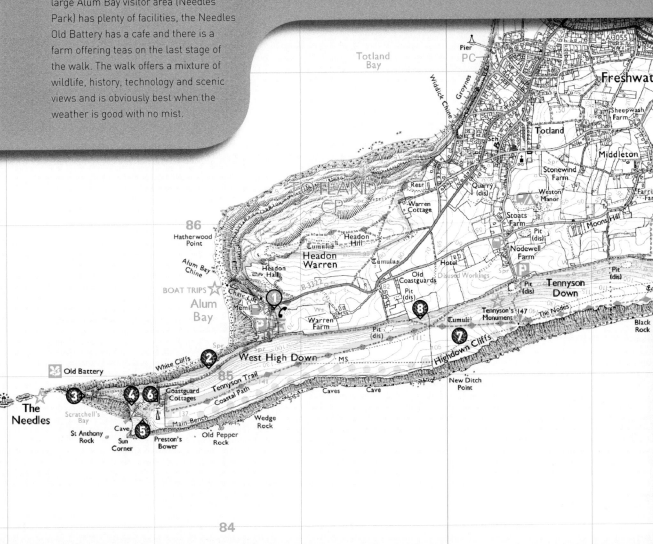

The Jurassic Coast
Lulworth Cove to Tyneham

HARD

ACCESS

4½ MILES

4:30

①

LULWORTH COVE

The rocks of Lulworth were formed between 150 and 65 million years ago. Major movements in the Earth's crust caused the strata to be uplifted and folded into the amazing structures that we see today. Erosion of softer strata has created the sheltered cove at Lulworth, protected on the seaward side by more resistant Portland limestone. At the back of the cove there are chalk cliffs, which are also more resistant, only collapsing when undercut by the sea.

With stunning views of magnificent cliffs and unspoiled beaches, this walk takes in some of the most dramatic scenery along Dorset's Jurassic Coast, as well as a wealth of unusual lime-loving plants, butterflies, other wildlife and a fossil forest.

This walk takes you across part of the Lulworth Ranges. Bindon Range has been used since 1916 as a tank firing area, while the Tyneham Valley was requisitioned in the Second World War for battlefield practice. Before you start the walk, you may wish to visit the Lulworth Heritage Centre, where there are excellent information boards and displays about local geology and natural history.

① Start the walk at the car park by the Lulworth Heritage Centre in **Lulworth Cove** and proceed across the pebbly beach towards the eastern end of the cove. Climb the steep steps at the far side of the cove and follow the cliff-top path signed to Peplar's Point, from where you can admire the clear turquoise blue water with swathes of delicate floating seaweed. Here a stone slab monument commemorates the life of Sir George Lionel Peplar, a tenant of Little Bindon parish for 50 years. As the cliffs immediately beyond Peplar's Point are unstable, take a right turn back to the main path and proceed along the marked cliff-top path towards the fossil forest.

② When you reach the flagpole, go through the gate in the wire fence, proceed

to the right and climb down some steep wooden steps. A remarkable 135 million-year-old fossil forest is located on a ledge some 80 feet above the sea. The forest is best observed from above, but it is possible to climb down onto the rocky ledge to examine the **fossilised tree stumps** at close quarters. The trunks are up to 6 feet in diameter. Their strange hollow shapes are the result of the interior of the tree rotting away, leaving only the hard fossilised outer parts. Longer, coffin-shaped holes represent the trunks of fallen trees. Adjacent rocks contain evidence of freshwater fossils and holes once occupied by salt crystals deposited from an evaporating lagoon during hot, dry summers. Some of the fallen blocks contain abundant rounded pellets (peloids), formed from the droppings of organisms that fed on the sediment that formed the blocks.

Climb back up the steps, turn right onto the cliff-top path and proceed about a mile towards Mupe Bay. Along the way and to the left you will pass the ruined chapel at Little Bindon. The cliff-top path at this point is carpeted with wildflowers – yellow trefoil, vetch, cowslip, wild parsnip, blue harebell, milkwort, pink and mauve thistle, wild thyme, scabious and sciancywort. Butterflies abound, and on a fine summer's day you might see small skippers, marbled whites and the dark green fritillary, all of which need chalk-loving plants for their life cycle. Birdlife is plentiful, with skylarks, meadow pippits and the occasional kestrel soaring in the skies, while gulls, kittiwakes, cormorants, guillemots and razorbills hug the cliffs.

③ Continue along the path to the ruined concrete wartime pillbox. This is a good opportunity to rest at the picnic table and admire the view of **Mupe Rocks** dipping dramatically into the sea. Although tempting, you should not try to clamber down to the pebbly cove here, as there is no safe access. Proceed to the left past the pillbox and follow the path along the top of the cliff to Mupe Bay. Along the way the path is particularly rich in gorse and wild blackberry.

④ At Mupe Bay, close to a picnic table on the cliff-top, there are steps down to the beach. As you descend the steps, note the abundance of wild sweet pea and, again, blackberry. Although similar to Lulworth Cove in appearance, Mupe Bay is far less crowded and is a very good place to stop for a swim or a picnic, although there are no public facilities here. The beach is made up of flint pebbles, many containing cavities with the traces of fossil sponges, around which the flint was formed. The flint pebbles have been eroded out from the chalk cliffs and at the far end of the beach fallen blocks of chalk may be found with flint pebbles still intact. Also look out for pieces of **black coal** (fossilised wood), which is preserved in the cliffs closer to the steps. Some of these cliffs are unstable as a result of landslip, so don't attempt to climb them.

⑤ Return to the top of the steps. At this point you may wish to take the alternative circular route back to Lulworth Cove by following the path to the right, over the stile and along the gravelled path, which takes you up to Radar Hill, from where there are magnificent views of Lulworth Cove and

FOSSIL FOREST

The Jurassic fossil forest contains a superb collection of fossil tree stumps preserved when a forest flooded some 135 million years ago. The trees seem to have been an early variety of cypress or juniper. Fallen blocks to the east of the ledge contain freshwater fossils and cubic crystals replace sea salt, which was formed by the evaporation of the shallow sea that drowned the forest as the sea level rose.

inland across to Lulworth Camp. The path then descends to Lulworth Cove, emerging at the cafe close to the wooden jetty.

If you wish to continue on the linear walk, take the steep path up Bindon Hill, bearing left at the top towards Cockpit Head. This is extremely arduous and not recommended for casual walkers. Follow the narrow cliff-edge path which descends precipitously to Arish Mell. The beach at Arish Mell is a danger area and is closed to the public.

6 From Arish Mell take the steep path eastwards up to Flower's Barrow. This Iron Age fort covers about 15 acres and lies on the cliff-edge known as Rings Hill. About one third of the hill fort has collapsed into the sea as a result of erosion before 1774. Three sides of the fort remain, with two banks and ditches and an original entrance to the south-east. Occupation platforms are visible, as are quarry ditches resulting from digging out material to build the ramparts. No archaeological excavation has taken place here, so the exact age of the fort is not known. However, it is thought to date from the fourth century BC.

From Flower's Barrow the path meets the old Dorset Coastal Ridgeway, which descends steeply into the Tyneham Valley above **Worbarrow Bay**.

7 Follow the path along the cliffs towards the eastern end of Worbarrow Bay. Here the path meets the gravelled track up to Tyneham village. At this point, have a look at

the display boards which explain the history of the Tyneham Valley and the people who lived there. Also look at the **Tett turret**, a small, rather ugly steel construction set into the ground and used by the Home Guard for anti-aircraft and machine gun defence during the Second World War.

8 Follow the gravelled path for about a mile towards Tyneham. The path follows the edge of Tyneham Gwyle (pronounced 'goyle'), a wooded area that was once maintained as hedgerow to restrain livestock. Since Tyneham was requisitioned by the Ministry of Defence (MOD), the unattended hedgerows have grown into small trees. As the trunks are now embedded with shrapnel they cannot be cut by chainsaw and so the woodland has developed into a haven for wildlife. As the path approaches Tyneham, there is a left-hand turn signposted to the woodland walk. Before taking this turning, spare a few minutes to look at the display board some 50 yards further up the gravelled path. Return to the woodland walk, which takes you through a small part of the wood, through shady bowers full of birdsong, to emerge at the car park at the far end.

This marks the end of the walk, but before departing you should take the opportunity to look around the small ruined village of **Tyneham**, uninhabited since December 1943. Tyneham House, the home of the Bond family for centuries, was of considerable antiquarian interest until its demolition a few years ago. Few buildings remain, but the exhibition in the church is well worth a visit. A useful leaflet documenting the history of the various village buildings is available for purchase in the church.

The Jurassic Coast

ADVICE

The walk is not suitable for wheelchairs or buggies. Most of the walk is on an MOD firing range. This, along with Tyneham village, has limited opening times, but is open almost every weekend of the year and every weekday in August. Walkers must stay on the designated paths due to the danger of unexploded shells and dogs should be kept under control. There are numerous shops and catering outlets at Lulworth Cove, and toilet facilities at both ends of the walk.

PARKING

Car parking is available at both Lulworth Heritage Centre and Tyneham.

START

The car park at the Lulworth Heritage Centre in Lulworth Cove is your starting point. The walk is linear and there is no public transport between Lulworth and Tyneham, although there is an alternative route if you wish to return to the start on foot, rather than proceeding all the way to Tyneham village.

CONTACT DETAILS

The Lulworth Heritage Centre, Lulworth Castle, East Lulworth, Dorset BH20 5QS
t: 01929 400352
e: webmaster@lulworth.com
w: lulworth.com

Ordnance Survey Explorer Map number OL15
© Crown Copyright 2008

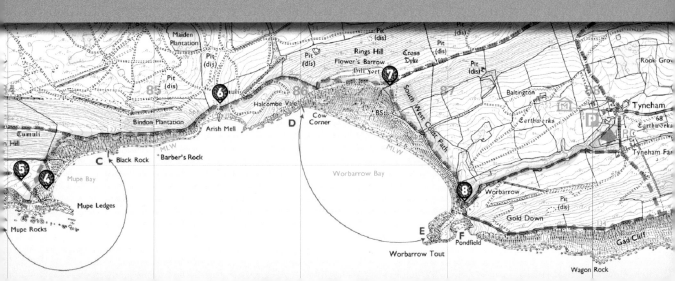

Lyme Regis
The pearl of Dorset

EASY

ACCESS

2½ MILES

2:30

This is a really good walk for families with young children. It is not too far and there are many points of interest along the way, plus several excellent beaches suitable for a picnic or fossil collecting.

1

Start the walk at the Charmouth Road car park and proceed down the hill. Lyme Regis was once a prosperous fishing port, exporting wool from the rich towns of Somerset. In the early 19th century the prominent limestone cliffs were quarried (mostly for cement which would set underwater) exposing large areas for **fossil hunting**. The town became very popular at this time when major fossil discoveries, including marine reptiles, were made by Mary Anning, the notable female geologist who became one of the town's most famous residents. Mary's work had a huge impact on life in Victorian England, when the Earth was believed to be only 6000 years old; Mary's fossil discoveries caused scientists to question whether the Earth was in fact much older. Lyme Regis also features in Jane Austen's novel *Persuasion*. More recently it became the spectacular backdrop for the film based upon John Fowles' novel *The French Lieutenant's Woman*.

2 After about 500 yards you will reach the church of **St Michael the Archangel**. Call in here to see Mary Anning's gravestone in the churchyard. Mary was born in 1799

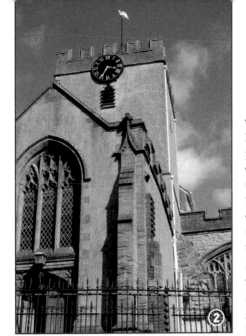

and was one of at least ten children born to Richard and Molly Anning, although only she and her brother Joseph survived childhood, the rest of their siblings dying before the age of five, probably from measles or smallpox. Mary died of breast cancer at the age of 47 and is buried with her brother Joseph (who died two years later) and three other siblings. There is also a stained glass window in the church dedicated to Mary's memory. The window is located to the left beyond a small set of steps and looks out over Mary's grave.

On the north wall of the church hangs an interesting mediaeval tapestry thought to depict the marriage of Henry VII to Elizabeth of York. The tapestry was first presented to the church in 1886 where it hung for a number of years in a rather sorry state of repair. Thereafter followed a brief sojourn of some 20 years at Trerice in Cornwall, and conservation work at Hampton Court Palace, until the tapestry was finally returned to the church in 1996.

3 Leave the church and continue walking down to the bottom of the hill until you reach Bridge Street. After a very short distance along Bridge Street you will pass the Guildhall on the left. The Guildhall has stood since Elizabethan times, when George Somers, the discoverer of Bermuda, was mayor and MP for Lyme Regis. Next to the Guildhall is the Lyme Regis Philpot Museum,

which stands on the site of Mary Anning's house and fossil shop (marked by a blue plaque on the wall). Mary Anning's family were desperately poor and supplemented their income by collecting fossils from the beach and selling them. The house and shop were demolished in 1889 to make way for the museum. This small museum was built between 1900 and 1901 by Thomas Philpot and from the beginning has been run by volunteers. It houses displays of maritime and domestic objects as well as fossils (in the geological galleries) and literary memorabilia from Jane Austen to John Fowles (in the writers' gallery). Just beyond the museum, on the right hand side of the road, is the **Lyme Fossil Shop**, which stocks a huge variety of fossils and minerals from around the world and is well worth a visit.

4 At the sea front turn left and walk out along Gun Cliff Walk, which, in earlier times, was lined with cannons. At the end of Gun Cliff Walk you can enjoy wonderful views of the Jurassic Coast to the east of Lyme Regis, including Charmouth and **Golden Cap**. If the tide is out you may wish to clamber down onto the rocky foreshore of Broad Ledge and walk out past Church Cliffs towards the grey cliffs of Black Ven. This was the fossil hunting ground of Mary Anning and here you should easily be able to spot large whorled ammonites, shells, burrows and other fossils exposed on the rock platform and in large boulders. Take care over the slippery seaweed and keep a watch on the tide to avoid being cut off at Church Cliffs.

FOSSIL HUNTING

Fossils are mostly found on the beach at low tide, either on the rock ledges or in large boulders, but they can also be found in landslips and occasionally in the cliffs. Please note, however, that as Lyme Regis is part of the Jurassic Heritage Coast, hammering of the cliffs is not permitted.

THE COBB

The great curving stone jetty of the Cobb protects the town from the ravages of the sea and creates one of the oldest artificial harbours in England. Dating back to the 13th century, it was probably originally wooden though displays many repairs and extensions since that time.

THE UNDERCLIFF

Stretching some 7 miles to the west of Lyme Regis, the Undercliff has been created by years of extensive landslips. It is now an unspoilt haven for a wide diversity of animals and plants, including self-sown ash and field maple, spindle, bramble, wayfaring tree, madder, clematis and everlasting pea. Springs and ponds contain giant horsetail, sedges and common reed.

⑤ Returning to Gun Cliff Walk proceed 500 yards along Marine Parade. About halfway along, a house to the right contains excellent ammonite fossils set into its front wall. The main sandy beach on the left is safe for swimming and an ideal place to stop for a picnic. Continue until you reach the picturesque **harbour**, where there are a few shops and catering outlets, and boat trips are available. From here walk out onto the Cobb at the far end of which is a small marine aquarium.

⑥ Return along the Cobb and at the harbour turn left along the beach road lined with yachts, caravans, chalets and **beach huts**. Here there were once kilns in which bricks were fired made from the local Blue Lias clay. At the end of the road, if you have suitably strong footwear take the opportunity to stroll along the pebbly beach in search of more fossils amongst the slumped debris at the base of the cliffs. At the far end of the beach a rock ledge and large fallen blocks reveal more large ammonite fossils. Above the cliffs is a vast stretch of land-slipped material known as the Undercliff. This is the place where Charles Smithson foraged for fossils and where he met Sarah Woodruffe in several romantic encounters in the film *The French Lieutenant's Woman*. The South West Coast Path passes through the Undercliff and can be followed all the way from Lyme Regis to Axmouth.

⑦ Return to the harbour and turn to the left up Cobb Road, past Holmbush car park, turning right at the top of the hill into Pound Street. This is a steep hill and a gentler alternative is to take the path through the **gardens** behind Marine Parade, where there are some colourful displays of coastal plants and flowers. Here there is also a Crazy Golf course from which you can catch a spectacular view of the harbour and the Cobb.

⑧ Proceed downhill along Pound Street, crossing Bridge Street into Sherborne Lane. Turn right into Coombe Street. On the left you will pass an elegant building housing the Dinosaurland Fossil Museum and shop. Further down the hill to the right you will find the entrance to Mill Lane and the Town Mill, which has existed since mediaeval times.

Here there is a working watermill, which operates daily (though it's worth checking the opening times), shop, bistro, art gallery and pottery. There is also a **small herb garden** planted to a 17th-century design. Leaving the Town Mill turn left and then immediately right along Monmouth Street towards the church. Turn left into Church Street and return uphill to the Charmouth Road car park, which marks the end of the walk.

Lyme Regis

ADVICE

Most of the walk is suitable for wheelchairs or baby buggies, though there is no wheelchair or buggy access either onto the beach to the west of the Cobb (beyond the beach huts) or onto Broad Ledge, Church Cliffs and Black Ven to the east. Close supervision of children is recommended near the base of the cliffs. The route is accessible all year round, but check tide tables if you intend to spend some time fossil hunting around Broad Ledge and Church Cliffs as these are cut off at high tide. There is a dog ban on the beach from 1 May to 30 September, but dogs are allowed on a lead at all times along Marine Parade and elsewhere.

PARKING

There are plenty of pay-and-display car parks in the town. The walk starts at the Charmouth Road car park, but ample parking is also available at Holmbush car park off Pound Street; there are other car parks nearer the beach, but these are short-stay only and tend to be fairly busy.

START

The walk starts and ends at the Charmouth Road car park to the east of the town.

CONTACT DETAILS

Tourist information Office, Guildhall Cottage, Church Street, Lyme Regis, Dorset DT7 3BS
t: 01297 442138
e: lymeregis.tic@westdorset-dc.gov.uk
w: lymeregistourism.co.uk

Tide tables online at www.bbc.co.uk/weather/coast/tides/southwest.shtml

Ordnance Survey Explorer Map number 116
© Crown Copyright 2008

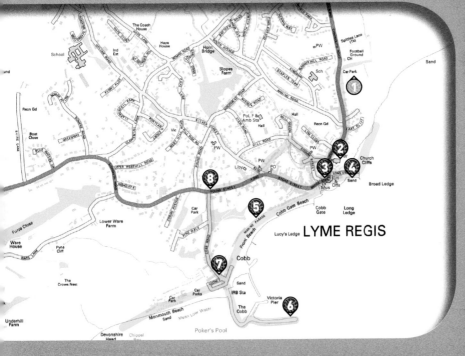

LYME REGIS

Devon's South Hams
The Warren to Noss Mayo

MEDIUM

ACCESS

4½ MILES

2:00

On this walk along the coastline of Devon's South Hams, from the Warren to Noss Mayo, you'll enjoy excellent views of the western approaches to the English Channel and there are several grassy areas where you can rest or picnic.

The Warren and Noss Mayo were once part of the Membland Estate, owned in the late 19th century by the first Lord Revelstoke, Edward Baring. The foreshore between Mouthstone Point and the Foot Ferry is designated a Site of Special Scientific Interest (SSSI) and the majority of the route takes you through an area of outstanding natural beauty.

1 Take the path from the Warren car park through a field and gate to meet the coastal path. Turn right and walk west. Follow the path round to the right, above a steep drop bounded by a dry stone wall. You will also see remnants of older walls once used to enclose rabbits, which were farmed for their skins and meat, hence the name the Warren. The immediate area is grazed by sheep, but bracken, gorse, elder and hawthorn grow unhindered, although as a result of the strong southwesterly prevailing winds, the elder and

hawthorn trees have low habits. If you look east and down, you can see **Blackstone Point** and it's well worth a short detour for the wonderful panoramic view, particularly west to Plymouth Sound and beyond to Cornwall.

2 Continuing towards Noss Mayo, the path passes immediately in front of Warren Cottage, once used as lunch stop for Lord Revelstoke and his sightseeing guests. Growing adjacent to this stretch of path are garden species such as mint, fuchsia and honeysuckle, but as you ascend a slight incline you'll note an increasing variety of flowering plants. Thistle becomes more frequent as you progress, and mullein and foxglove make their first appearance here as well. Bramble and gorse dominate the steep slopes down to the cliff-edge, while smooth hawk's-beard commonly grows at the path edge. However, occasionally you can also find clumps of **wild thyme**, pink campion, heather, scarlet pimpernel, sheep's-bit and the fairly rare sea campion growing beside

the wire fence that protects them from the sheep and cattle that routinely graze here.

3 After Warren Cottage the path is more deeply cut into the bedrock and you can get a good look at the slate as you pass above several cove inlets and protruding wave-swept **rock platforms** that make for a craggy irregular foreshore. The more level wave platforms are favoured for rod and line fishing, but care needs to be taken as most are submerged at high tide. During the summer, bass, garfish and mackerel may be caught here, and pollock and wrass all year round.

4 Once the next bend has been rounded the **Great Mew stone** – mew being the old name for gull – is now visible in the distance, with Cornwall as a backdrop. Continuing on, areas of outcropping rock become more pronounced and provide good examples of the gently dipping brown and grey, heavily cleaved slate, with its regular, elongated gaps and, more rarely, green and grey fine-grained sandstone. Areas which were once quarried are also evident, but they have now been colonised by bramble and gorse. The quarried rock was used locally in the construction of cottages and more prestigious buildings such as Membland Hall, although this was demolished in 1945. However, the artisan dwellings in Noss Mayo were generally constructed of cob – sand, clay, straw and water.

5 At Gara Point the path veers right and presents the walker with an uninterrupted view of Wembury Bay. As you pass the rock outcrop known as Gapmouth Rock, the mouth of the **River Yealm** lies immediately in front, with Wembury beach beyond. At the

ST PETER'S CHURCH

Built thanks to the patronage of Edward Baring, Lord Revelstoke, and designed by J.P. St Aubyn, with its organic carved oak panels the interior of the 19th-century St Peter's Revelstoke is a fine example of the Arts and Crafts Movement style. The medieval tradition of painted, stencilled walls was also renewed and the end result is very pleasing and reminiscent of William Morris's work. The exterior walls are mostly constructed of slate taken from various local quarries.

CHOLERA EPIDEMIC

In 1849 there was a cholera outbreak, primarily in Noss Mayo. At that time the exposed population was just over 600. Of these, 200 were infected with the disease and over 50 died. In some cases whole families perished.

DARTMOUTH SLATES

The rocks known as the Dartmouth slates were formed in the Lower Devonian period, roughly 380 million years ago, and are some of the oldest rocks in the south-west. Volcanic activity via fractures in the then sea floor produced the dolerite dykes now exposed in the faces of some cliffs. A major episode of rock deformation known as the Variscan Orogeny (mountain-building) event distorted the slates and accounts for the dipping and cleavage.

next right-hand bend there is a short detour that takes you down to Greylake Cove, where it may be possible to distinguish a dolerite dyke standing slightly proud of the cleaved dark brown slate either side of it. Much of the stone here has been removed by quarrying, because it is very hard and therefore a durable building material.

6 Returning to the main path you head inland following the southern side of the River Yealm, down to the National Trust's **Brakehill Plantation**. This is a mixed wood plantation comprising rhododendron, oak, hazel, elder, sycamore and beech. The undergrowth comprises various ferns, as well as bracken, bramble, nettle and pink campion. On leaving Brakehill Plantation the path enters open ground and splits to pass either side of the old Coastguard Cottages, before entering Passage Woods. The lower, left-hand track takes you down to Cellar Beach and then on to Battery Cottage, but the paths converge after a short distance. A few yards further on is the landing stage for the summer-only foot ferry to Warren Point. The woodland canopy continues into Ferry Woods, which skirt the River Yealm, and then the path bears right into Newton Creek.

7 On leaving Ferry Woods the route enters the village of Noss Mayo. The public road narrows here and is bordered by well cared for, pretty cottages and gardens. The road can be very busy in the summer, especially as you turn right into the small tidal inlet of **Noss Creek**. This road takes you behind the Ship Inn and on to the Village Hall and Tilly Institute.

Noss Mayo was first referred to in the 13th century, but by its original name of La Nasse de Matthieu, and was owned by Matthieu Fitz Herbert, who was of Norman origin. Once a thriving farming community, Noss Mayo later developed into a fishing village and the name 'Cellar', as in Cellar Beach, referred to a fish and haul equipment store. Noss Mayo is now a favoured holiday destination, especially for sailing enthusiasts. The village extends to Bridge End, its eastern limit with Newton Ferrers, and has two pubs, the Ship Inn and the Swan.

8 Before continuing your journey there is a short detour that you can take. Proceed past the village hall to your right and take the second turning on the left, round the sharp left-hand bend into Revelstoke Road. The 19th-

century church of St Peter's Revelstoke is at the end of the road and is entered via a **lynch gate**.

To return, retrace your steps as far as the sharp bend, then go straight on into Hannaford Road (not signed) with its tennis courts, play area and bus stop. Continue to the end where the road ceases and the footpath begins. The footpath lies within a steep-sided valley, out of which there is a relatively steep climb at its southern end. Continue past a few houses on the right and up the uneven track to the main road. Turn left and the entrance to the Warren car park is a few yards on the right.

Devon's South Hams

ADVICE

The majority of this route passes through National Trust coastal path and woodland. The route is open all year, but is not accessible for wheelchairs and buggies. Dogs are welcome, but must be on a lead when sheep and cattle are present. There are toilets in Noss Creek, near the Ship Inn.

If you're travelling by public transport, you can do the walk in reverse, as the number 94 bus route serves Noss Mayo from Plymouth.

PARKING

Follow signs from the A379 Plymouth to Kingsbridge Road and turn right at Flete Western Lodge, about 4 miles from Yealmpton, following signs for Battisborough Cross. From here follow signs for Stoke Beach. At the Stoke Cross junction go straight on to Netton. The Warren car park is on the left about three-quarters of a mile from Netton Farm. It's National Trust property and there is a donation box. There is very little parking in Noss Mayo, especially during the holiday season.

START

This walk begins at the Warren car park.

CONTACT DETAILS

Kingsbridge Information Centre, The Quay, Kingsbridge, South Devon TQ7 1HS
t: **01548 853195**
f: **01548 854189**
e: **advice@kingsbridgeinfo.co.uk**
w: **kingsbridgeinfo.co.uk**

Ordnance Survey Explorer Map number OL20
© Crown Copyright 2008

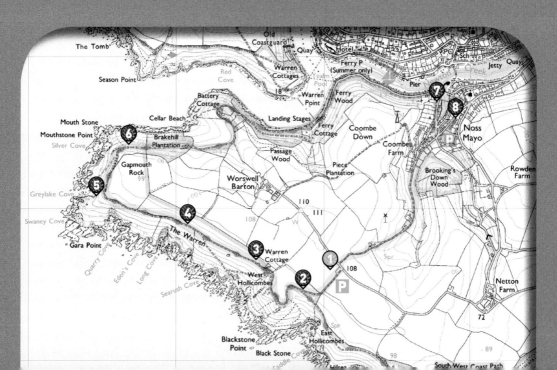

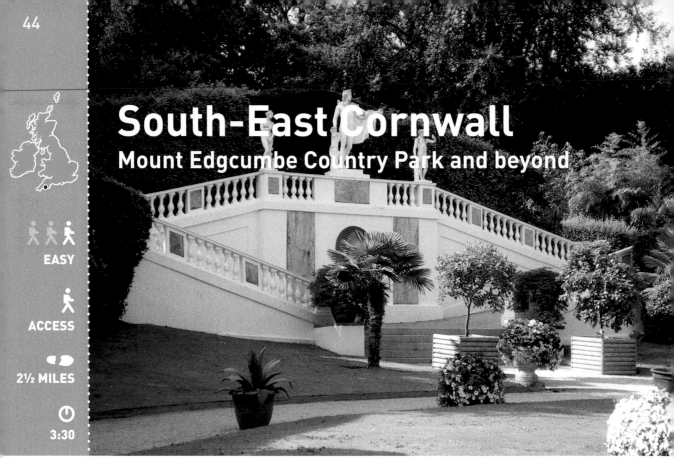

South-East Cornwall
Mount Edgcumbe Country Park and beyond

👣👣👣
EASY

🚶
ACCESS

👣👣
2½ MILES

⏲
3:30

This circular walk passes through the formal grandeur of Mount Edgcumbe Country Park and on to the wilder coastal path beyond.

. . . .

MOUNT EDGCUMBE COUNTRY PARK

The park is a gardener's delight at all times of the year, but especially so when the multitude of camellias are in bloom in May and June. Within the park there is an eco-bus service and after the walk it is a leisurely way to see the gardens. It is free, but donations are welcome. The house and country park are now jointly managed by Cornwall County Council and Plymouth City Council.

1 The walk starts as you enter the main gates of the park, and up and ahead of you stands Mount Edgcumbe House. To your right is the visitor centre and shop (free guides available) and to the left the eco-bus stop. Bear left towards a set of turreted gates, pass through these (toilets to the left) towards the orangery, which is now a restaurant and borders one side of the Italian garden. This garden was laid out sometime around 1785 and replaced the old farm. Circumvent the orangery, bearing left and follow the path through an opening in the hedge. This opens out into managed parkland. Keep to the far left, following the Tamar River foreshore towards the Block House built around 1540.

At Wilderness Point to the left is the Earl's Battery (*c.* 1747) and the **Garden Battery**, a later addition (*c.* 1863). Using the original compliment of 21 guns the Earl would herald

the seaward arrival of his guests. How many guns were fired depended on the importance of the guest: a king, for example, got the full complement. The three guns present today were taken from a captured French ship. The Battery offers the walker extensive views of Plymouth, Drake's Island and Royal Naval vessels anchored at the Devonport Dock Yard. Take a short detour here and descend the steps just before the Garden Battery to the rocky foreshore. Here you will get a closer view of the only example of Plymouth Limestone (Middle Devonian age, 394–383 million years) present in Cornwall. Retrace your steps and proceed past the block house and on through a gate that leads to the bowling green that borders a pebbled beach.

2 Continuing on, you arrive at **Barn Pool**, a stony beach and a favourite spot for picnics and paddling. Barn Pool is a sheltered, deep anchorage and from here the *Beagle* weighed anchor on 27 December 1831 and carried Charles Darwin on his famous five-year voyage. There is now a choice of two routes. If using the path above Barn Pool beach (the return route is signposted here), which is suitable for wheelchair and buggy users, follow the signs for the amphitheatre. This is a wide, hard-surfaced, but slightly inclined track that passes through mixed woodland before opening out in front of the small lake. There are various tracks off to the right, before the lake, that connect with return routes to the main entrance gate. There are several examples en route of 'folly'-type wells built of locally quarried stone, one of which is situated on this section of path.

3 Alternatively, for those on foot it is possible, at low tide, to walk the length of Barn Pool as far as the protruding Raven's

Cliffs. Take care, however, of the tides and the beach stones, which become slippery underfoot as the carpet of green algae increases. You may be able to identify the different types of pebble found here, e.g. pumice, Staddon Grit sandstone and bivalve bored Plymouth limestone. As you proceed along the beach you will observe sea anemones, barnacles and common limpets attached to the lower margins of rocks submerged at high tide.

At the far end of Barn Pool there are some very **interesting rock features**, which are especially visible within the gullies. It is advisable, however, to wear a hard hat when entering these gullies as the cliff faces are unstable (see the exposed tree roots above). The gullies are probably fault zones and as such have severely weakened the rocks. These structural weaknesses have enhanced erosion by wave action, which has led to the instability. The rocks outcropping along this section belong to the Saltash Formation, which traverses the Devonian and Carboniferous periods (390–365 million years). The rocks show several interesting sedimentary features and include flowing intermingling 'fingers' of black and grey slate, and, in discrete zones, oxidised inclusions within the fine-grained matrix of the grey slates. On reaching Raven's Cliffs retrace your steps back to the main path in front of the small lake.

4 As you reach the lake you will see immediately ahead of you the **temple dedicated to Milton** (*c.* 1755) with an inscription from *Paradise Lost*. After this point

REGIONAL GEOLOGY

The south-west is geographically divided into fault-bounded basins trending east-west, each comprising different rock sequences. Mount Edgcumbe is juxtaposed between the Looe and South Devon basins, and the former holds the oldest exposed rocks in Cornwall, up to 395 million years old. The sediments are a sequence of mudstones, siltstones and sandstones that denote deepening water in a reefal setting. There are also discrete basaltic intrusions and volcanics. In east Cornwall the South Devon Basin comprises, similar sequences but the sandstones are relatively less common.

⑤

CREMYLL FERRY

This ferry crossing is one of the oldest on the Tamar and was first mentioned in 1204. Traveller Celia Fiennes made this crossing in 1695 and commented in her memoir, *Through England on a Side-Saddle*, that she found the journey to be 'very hazardous'. She added her own 'men' to the five rowers, but despite this the journey took 1½ hours, including a 15-minute forced standstill where three tides converge. Today's crossing takes less than ten minutes.

MOUNT EDGCUMBE HOUSE

The current house is a faithful rebuild of the original, which was severely damaged by fire in 1941. Of the original Staddon grit house, built between 1547 and 1550, the Tudor walls and 18th-century towers were salvaged, around which construction continued between 1958 and 1964.

wheelchair users and probably buggies will find the path inaccessible and will need to retrace their path back to the picnic area in front of Barn Pool. From here follow the left-hand track and then fork right. Proceed down the Avenue back to the Cremyll entrance.

For those continuing onward, pass to the left of the temple and enter mixed woodland that supports a large variety of trees. The path narrows here and gets steeper as it passes above the Raven's Cliffs and Ravenness Point. Proceed through a gate and above on your right a folly in the style of a ruined tower comes into view (*c.* 1747). The folly was constructed using stone from the medieval churches of St Laurence and St George in East Stonehouse. On the other side of the folly is the Earl's Drive, from which there is a better view of the folly. The Earl's Drive dates from 1788 and was extended to Penlee Point in 1833.

⑤ Continue uphill a short distance until the path levels off and take the second fork, which leads you down to the undulating coastal path and eventually to a bolder-strewn beach. Take care along this path as underfoot several tree roots and the occasional cable are exposed. To your right you will see Lady Emma's Cottage, quite on its own in the wood. This mock Tudor cottage replaced a thatched beechwood cottage destroyed by fire in 1882. To your left you will see numerous navigating buoys warning ships and boats of the submerged rocky outcrop known as the Bridge, which can be traced back to Drake's Island. Following the direction arrows proceed across a boardwalk over a small pond bordered by boggy ground colonised by mint, pink campion and iris. Just past here is a small beach with an excellent view of **Drake's Island** through the tree foliage.

⑥ Just before Redding Point, head slightly inland and uphill along a single track to a fenced off section of pathway. Bear right here and take the wooden steps up to the path above. Here another landslip makes it necessary to take a further stepped detour

that connects with either the forward coastal path to Fort Picklecombe or the alternative return paths, either through the Deer Park or along the **Earl's Drive** to the house. There is a dextral (movement to the right) strike-slip fault heading north-west from Redding Point and this marks the contact between the Saltash and Staddon Grit Formations. Movement along this fault accounts for some of the cliff instability and, as with all faults, constitutes a zone of weakness that facilitates cliff collapse by wave action.

⑦ Should you wish to head for home now, retrace your steps and take the Earl's Drive, which passes through the mixed woodland with camellias. On leaving this woodland canopy you enter open grassland with the ruined **tower folly** to the right. Above the folly there is a seat bench from which to view the vast expanse of Plymouth Sound, Dartmoor and the Lee Moor china clay works in the distance. From here the path bears left and, passing through a gate, you once again enter woodland. This wood comprises numerous collections of camellias affiliated to various regions of the world, e.g. Australia, New Zealand and the USA. Continue on into open parkland once again and, when meeting up with various connecting paths,

⑥

Fort Picklecombe, passing **the arch** (*c.* 1760), which was relocated here from Stonehouse. Continue on for a short distance and pass above Fort Picklecombe, now converted into apartments. You meet a small road and by bearing right you pass Maker Farm on your left and eventually connect with the main road between Millbrook and Cremyll. At this junction turn right. Head down this road for a short distance and re-enter Mount Edgcumbe between the imposing gate pillars. In 1763, Maker Church, which appears on your right, was a naval signal station. Proceed down the Avenue to the main entrance.

bear right towards Mount Edgcumbe house, passing behind it. Follow signs for Cremyll, heading downhill along the Avenue.

8 If you want to continue your walk for a while, at the steps go on towards

South-East Cornwall

ADVICE

Parts of the walk are accessible to wheelchair users and buggies by using the numerous concreted tracks (courtesy of the US army). The exception is the lower section of path between Milton's Temple and Fort Picklecombe. The walk to Fort Picklecombe heads inland, because of landslips along the coast, which are clearly identified en route. Hard hats should be worn when approaching the cliff faces between Barn Pool and Redding Point.

PARKING

There are pay-and-display car parks at Cremyll and at the Country Park.

START

The start of the walk can be accessed either by car to Cremyll, or by foot ferry from Admiral's Hard, Stonehouse, Plymouth.

CONTACT DETAILS

Mount Edgcumbe Country Park, Cremyll, Torpoint, Cornwall PL10 1HZ
t: **01752 822236**
e: **mt.edgcumbe@plymouth.gov.uk**
w: **mountedgcumbe.gov.uk**

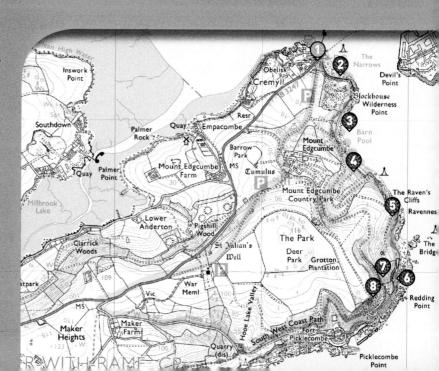

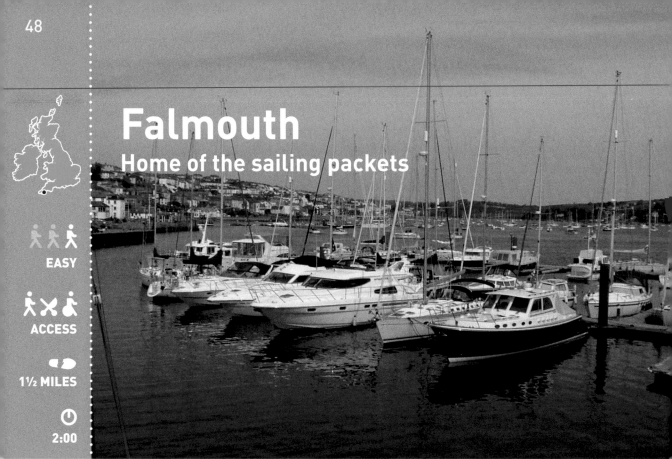

Falmouth
Home of the sailing packets

EASY

ACCESS

1½ MILES

2:00

THE PACKET SHIPS

By 1702 packet ships were sailing regularly to Lisbon in Portugal and to Barbados and Jamaica in the West Indies. New York was added in 1755, and by 1784 packets were serving Gibraltar and several other ports in the American colonies. Routes were opened up to South America and into the Mediterranean at the beginning of the 19th century. The mail for India was taken to Alexandria, where an overland link to the Red Sea was established in 1835.

One of the best natural harbours in the south-west of England, Falmouth's history has been closely tied to that of the sailing ship and for over 300 years has been the first port of call for ships returning on the prevailing westerly winds.

The walk begins in Event Square in front of Falmouth's newest attraction, the National Maritime Museum. Situated in a part of the town that also houses Port Pendennis – Peter De Savory's luxury apartment complex – upmarket coffee shops and restaurants, this area is evidence of a thriving new community.

You can also enjoy the tranquil views of the yachts on the water, along with Cornwall's largest working port – Falmouth Docks. This was the spot where thousands of people joined the world's media under blue

skies in February 2005 to welcome **Ellen MacArthur** and her 75-foot trimaran *B&Q* home after her 71-day record-breaking solo circumnavigation of the world.

Falmouth has played this role before. In 1969 it greeted Robin Knox-Johnston on the first solo voyage around the world and he was among the 8000-strong welcoming committee some 36 years later.

② Walk across the square away from the museum and you will see a strange pyramid-shaped obelisk. This is the **Killigrew monument**, erected in memory of one of the most important families in Falmouth, whose story is one of piracy, greed and ambition.

For as long as there are records of the town, the Killigrews were the power behind it. During the 16th and 17th centuries the family seemed to have it all. Peter Killigrew persuaded Charles II to make the town the Royal Mail Packet Station, where letters and gold bullion were sent from around the world, bringing wealth, power and influence to the town and the family.

But everything came to an end in the 18th century when Peter's son was killed in a duel. The only thing left was this monument. But however closely you look at it, you won't find any hint as to what it is really for as it is unmarked. Local gossip has it that inside the obelisk are two glass bottles, last seen when it was moved in the 1830s. But to this day, no one knows what the bottles contain.

③ Now head up the main road into town. Soon after Trago Mills you will see a sign for Custom House Quay on a wall in front of you. Turn right down the hill to the **quayside**. Look across the water at Trefusis Point – it's a dangerous place. It was here that the worse accident in Falmouth harbour occurred in 1814. A government transport ship, the *Queen*, bringing soldiers and their families back from the Peninsular War, was driven into the rocks by an easterly force 12 gale. Nearly 250 people drowned. According to legend a marine monster named Morgawr has been sighted here and on the night of 12 and 13 July 1940 several bombs fell here for no apparent reason.

④ Turn back on yourself and walk around the dock, keeping the water on your right. Because of its position Falmouth was chosen as the base for a scheduled government route to Coruña in Spain in 1689. The mail came from London, first by horse and then by mail coach, and was transported on packet ships chartered by the Post Office. For the next 150 years Custom House Quay was the only place where the mail came in and out of the country.

The mail was not just made up of letters and postcards, the boats carried vital and often secret intelligence. The **packet ships** attracted a lot of attention from pirates and others keen to intercept them. They had to be fast – and they were. Small, two-masted brigs, they

THE GREATEST RAID OF ALL

An obsolete destroyer, HMS *Campbeltown*, was at the centre of the raid on St Nazaire. Packed with explosives, the ship landed the commandoes before being rammed into the lock gates and scuppered at around midnight. The crew set time-delay fuses before disembarking in the confusion. Despite the appalling casualties inflicted by the Germans, having done their jobs some of the sailors and a number of commandoes escaped. The ship eventually exploded at around 10.30 a.m. the following morning, killing 250 German soldiers and inflicting huge damage on the docks.

carried square sails, a long bowsprit, headsails and a large gaff mizzen-sail and could regularly outrun a lumbering man-of-war.

In its heyday, 3000 ships a year passed through Falmouth. But steam was its undoing. The steamships were faster and more reliable than the old sailing packets, and they had no problem reaching London whatever the wind. By 1850 the town's place as the start of the information superhighway was over.

5 Head up the ramp out of the car park and back onto the main road. Turn right. Walk down the street until you reach the Falmouth Arts Centre. This building, known locally as the Poly, began life as the Royal Cornwall Polytechnic. Founded in 1833 by members of the Fox family, prominent local Quakers, the Polytechnic aimed to promote the arts and sciences to workers in the family's shipping and foundry businesses.

They bought their own building for an annual exhibition of the arts and sciences. Exhibitions related particularly to Cornwall, but also featured all aspects of the arts, sciences and manufacturing, often highlighting the living conditions of workers in local industries like mining and fishing. The society played a prominent role in improving industrial life and in the development of local arts like photography. William IV became patron in 1835 and the society added 'Royal' to its name.

Today the centre presents a regular programme of drama, film and art.

6 Now head down the street again and turn right opposite St George's Arcade,

going down the ramp into the car park. Walk towards the railings and then left to Fish Strand Quay. Stop at St Nazaire memorial.

This memorial commemorates one of the most daring raids of the Second World War. On 26 March 1942, a modified warship carrying more than 600 men left Falmouth on a secret mission. The aim of Operation Chariot was to destroy the heavily fortified St Nazaire dock in northern France. This was the only place that the mighty German battleship *Turpitz* could go for repairs, so it was vital that the docks were destroyed.

The Allies realised that the dock was only vulnerable to a two-pronged attack. The idea was to land a force of men to demolish the port and ram the dock gates from the sea. Against the odds the daring mission was accomplished, but not without casualties. More than 160 men died and 200 were captured. Five Victoria Crosses were awarded.

7 Now head along the railings with the sea on your right until you reach the Battle of Trafalgar memorial. Today Fish Strand Quay houses a busy car park, but on 4 November 1805 it was the setting for an event of huge historical significance.

HMS *Pickle* appeared over the horizon, bringing with it one Lieutenant Lapenotiere. The ship anchored offshore at Pendennis Point and a longboat delivered Lapenotiere to land. The Lieutenant was on a mission. He had been instructed by Vice-Admiral Collingwood to deliver a dispatch to William Marsden, the Secretary of the Navy, at Admiralty House in London as soon as possible. The news he brought into Falmouth was momentous. There had been victory at Trafalgar, but at the cost of the death of Admiral Lord Nelson.

Lapenotiere ordered a 'post-chaise' to take him to London and after 36 hours, 21 changes of horse and 271 miles the determined Lieutenant delivered his important dispatch.

8 Now walk away from the memorial and back through the car park. Head up the ramp and turn right, back down the main street. Look out for **Bell's Court** on your left.

As you stand looking into the quiet square it's hard to believe the small lane was once the scene of a major revolt. If you had been standing in the same spot on 24 October 1810 you would have been surrounded by angry men.

There was mutiny in the air outside the package agent's office that day. The protestors were all sailors on the packet ships. They were angry that one of the customs men had taken possession of goods that the sailors had imported believing that it was one of the perks of the job. It was an explosive situation.

From the top of the steps outside the office, agent Christopher Saverland read the Riot Act to the angry sailors. The order stated...'[The men should] disperse themselves, and peaceably to depart to their habitations or to their lawful business, upon the pains contained in the Act made in the first year of King George for preventing tumultuous and riotous assemblies. God Save the King.'

As his words sounded there was a glint of metal behind him in the sunshine as the soldiers' bayonets were fixed. The local militia had come down to enforce the Act.

The crowd eventually dispersed... and it's all been rather quiet in Bell's Court since then.

If you turn around and head back along the high street you will eventually get back to your starting point.

200 YEARS ON

In 2005 a re-enactment marked the second centenary of Lieutenant Lapenotiere's actions after Trafalgar. Following a reception in the National Maritime Museum, a naval officer left by a specially built post-chaise for Truro on the first stage of the journey to London. The Ordnance Survey produced a commemorative map of the original route.

Falmouth

ADVICE

Most of the walk goes through the main streets of Falmouth where there is a one-way traffic system. The walk often leaves the streets behind and heads down to the quays and these are reached via quite steep hills. However, the walk can be done at own your pace.

PARKING

There are several pay-and-display car parks near the start of the walk at the National Maritime Museum. The car parks charge by the hour, but evenings and Sundays only cost £1 per visit.

START

The walk begins just minutes away from the railway station at the 'Trago' end of town.

CONTACT DETAILS

**Falmouth Tourist Information Centre,
11 Market Strand,
Prince of Wales Pier, Falmouth,
Cornwall TR11 3DF
t: 01326 312300
f: 01326 313457
e: info@falmouthtic.co.uk
w: acornishriver.co.uk**

Ordnance Survey Explorer Map number 103
© Crown Copyright 2008

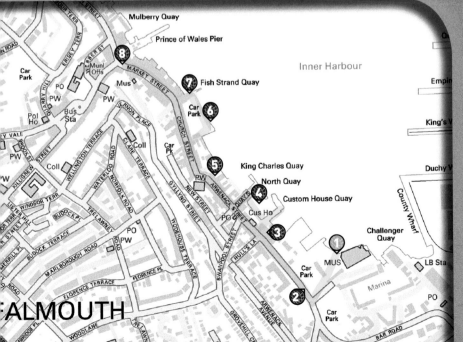

Tresco
Subtropical paradise

MEDIUM

ACCESS

5 MILES

4:00

8

INVASION

The Scillies have always been vulnerable to attack and were often seen as an easy target by foreign powers and pirates. The *Orkneyinga Saga*, which dates from the late 12th century, recounts the story of a Viking raid on the Scillies. However, sometimes would-be invaders encountered more resistance than they might have anticipated and, in 1209, the monks of Tresco Abbey are said to have beheaded 120 pirates in one afternoon.

Tresco is one of the most remote and idyllic of the British Isles. This circular walk around its coast takes in the wild and windswept moorland at its northern tip, as well as the famous subtropical Abbey Gardens to the south, with unspoiled sandy beaches, castles and stunning views to the other Isles of Scilly en route.

Tresco is one of the Isles of Scilly, an archipelago of 56 small islands that lie 28 miles south-west of Land's End. This walk starts at Carn Near Quay at Tresco's southern tip, then follows the coastal path clockwise through the village of New Grimsby and past Cromwell's Castle in the north-west, before cutting across the island's rugged northern peninsula and continuing down the east coast to the incredible Abbey Gardens.

1 Take Carn Near Road north from the quay to the heliport at Abbey Green.

Cross the green carefully (red lights at its entrances are lit when helicopters are about to take off or land) and take the road to the left that leads behind the heliport terminal buildings. A footpath, which soon forks left from this road, takes you briefly along the edge of the beautiful beach in **Appletree Bay**, from where there are wonderful views of the islands of St Agnes and Samson.

Continue north, soon rejoining the main coastal path. The long island across the water on your left is Bryher, another of the Scillies'

five inhabited islands and Tresco's twin. As the path winds right, the fortification known as Cromwell's Castle emerges in the distance, across Saffron Cove. At this point, in fair weather, the famous Bishop's Rock lighthouse can be seen in the distance to the south-west, between Samson and Bryher.

2 The northern end of Saffron Cove is marked with a green knoll called Plumb Island. Here the coastal path goes downhill, past a lake known as the Great Pool on the right, to join Abbey Drive. The lake was dug by medieval monks to provide a supply of freshwater fish to Tresco Abbey. Continue left along the signposted path which soon brings you to the settlement of New Grimsby. Here there is the Estate Office (which administers the amenities and holiday cottages on the island), various shops and public toilets. Continue along the surfaced road beside the shoreline. An old seaplane slipway can be seen to the left.

3 Make your way around New Grimsby Bay, past Tresco Gallery and the row of pretty cottages on your right, to New Grimsby Quay, which is the high-tide landing point for ferries from St Mary's and has public toilets. At this point the character of the walk changes markedly, as you enter the wilder, more rugged terrain of the north of the island. The paths here are narrow and steep in places, and therefore best avoided by those with mobility difficulties.

4 Continue along the dirt track behind the quay building, which soon curves right, where a bench provides more wonderful views of Bryher. The path continues up to Braydon Rock, which offers the best view of Cromwell's Castle. Make your way carefully down the other side of the rocks to where the dirt path continues along the coast, past Frenchman's Point, to Cromwell's Castle

itself. This 17th-century fortification, which is unstaffed, but maintained by English Heritage, is open all year round. Entry is free and visitors can inspect the battery of cannon in its lower courtyard and climb the steps to its rooftop look-outs, which provide commanding views along the length of New Grimsby Channel and Bryher beyond.

On leaving Cromwell's Castle, a signpost directs walkers steeply uphill towards **King Charles's Castle**, which was built in 1550 to guard against French or Spanish invasion. Although it is now a roofless ruin, most of the castle's walls and doorways are still fairly intact. From here there are magnificent views of the whole archipelago.

5 From King Charles's Castle, take the track that heads towards the northern tip of the island. As you approach Gun Hill you have a choice. In summer, when the purple heather is easily traversed, turn right and continue directly across the heathland, towards the white lighthouse on the uninhabited Round Island beyond **New Grimsby Sound**. However, in winter and wet weather it may be preferable to take the lower route, following the coastal path around the northern tip of the island.

The more adventurous may wish to explore Piper's Hole, a cave that extends about 200 feet into the cliff here. It opens out to a cavern that contains an impressive 30-foot-wide freshwater pool and was once a favourite haunt of smugglers. However, do take torches

AUGUSTUS SMITH

In 1834, the Hertfordshire banker Augustus Smith (1804–1872) leased the Scillies from the Duchy of Cornwall for £20,000, gave himself the title Lord Proprietor and made Tresco Abbey his home. During a governorship of more than 30 years, his development of the flower-producing and tourism industries helped to lift the islanders out of the poverty in which he had found them when he arrived. He funded a school-building programme and started the world-famous Tresco Abbey Gardens.

and appropriate precautions. You join the main route again at Gimble Porth.

6 Those taking the high route over the heather will, after about 450 yards, join a clearly visible path, as it descends towards the rock-strewn beach at **Gimble Porth**, with views towards Merchant's Point at the far side of the bay and the uninhabited island Northwethel to its left. The path continues over a sandy rise, soon arriving at the settlement of Old Grimsby. The first buildings you come to are part of the luxury Island Hotel, the main entrance of which is on your left. The hotel is open to non-residents and is another convenient refreshment stop.

Continue past the entrance to the Island Hotel and along the main path, with Old Grimsby beach on your left, until you come to Old Grimsby Quay, at which point the path bears sharply right and then forks. Those wishing to visit St Nicholas's Church should take a short detour to the right. The church is dedicated to St Nicholas, the patron saint of seafarers, and was built in 1882 in memory of Augustus Smith.

7 To continue the walk, take the left-hand fork, a sandy path that leads past a community centre and playing fields on the right, with views of the church beyond. A little further along, on the left, are some cottages and a path leading uphill to the Blockhouse, a fortification built in the mid-16th century, around the same time as King Charles's Castle, to protect Old Grimsby harbour and sound from invasion. It was held by the Royalists during the English Civil War. Take a detour to explore it, if you wish, but the walk proper continues along the main path, which forks again a little further on. Take the left-hand

route down Borough Road. To the left, there are sheltered fields, to the right, a dry-stone wall with a large drinking trough set into it. Continue along this leafy path, past Pentle House on your right, and skirt around the east bank of the Great Pool, along Penzance Road, until you reach **Tresco Abbey**.

In the 12th century, when sea levels were lower than they are today, the northern Isles of Scilly were joined together as one single island, whose religious estates Henry I granted to the Benedictine monks of Tavistock Abbey in Devon. At this time, Tresco Abbey, more properly a priory, was dedicated to St Nicholas. The Abbey is home to the Dorrien-Smith family, the descendants of Augustus Smith.

8 Continue along the path, past the main façade of the house on your right, to the entrance to the magnificent Abbey Gardens. This 12-acre haven, which is best visited during the spring and summer months, is uniquely positioned to take advantage of the North Atlantic Drift, that powerful ocean current which brings with it the warming effects of the Gulf Stream. Frosts are rare here and a barrier of trees, planted by Augustus Smith, protects the gardens from harsh northerly winds (prior to this there were no trees at all on the island), to produce sheltered conditions that are capable of sustaining subtropical plants and trees. Many species, such as the striking pink proteas, are unable to survive elsewhere in the British Isles.

The gardens are open daily between 10.00 a.m. and 4.00 p.m. (there is an admission charge), and it is worth spending at least a couple of hours exploring them. An area of the garden called Valhalla, near the southern entrance, houses a large collection of figureheads salvaged from the many ships wrecked around the Isles of Scilly over the years.

Tresco

ADVICE

Although the dirt tracks and slightly tricky rocks of Castle Down and Tregarthen Hill put this walk firmly in the 'medium' category, the route can be shortened and made very much easier by taking the road from New Grimsby to Old Grimsby, via the New Inn (another possible refreshment stop) and St Nicholas's Church, thereby avoiding the north of the island entirely.

TRANSPORT

There are regular flights to St Mary's, the largest of the Isles of Scilly, from Newquay (30 minutes) and Land's End (15 minutes) or you can take the helicopter from Land's End to St Mary's or Tresco. The ferry from Penzance to St Mary's takes around three hours. There are no cars on Tresco, although golf buggies can be booked from the Estates Office in New Grimsby for those who need them.

START

The walk starts at Carn Near Quay, because this is where visitors taking the ferry from St Mary's at low tide land. However, as the route is circular, those arriving at other quays or at the heliport, or staying elsewhere on the island, can pick up the route at the nearest convenient point and complete the circuit from there.

CONTACT DETAILS

**Hugh Town, St Mary's,
Isles of Scilly TR21 0LL**
f: 01720 423782
e: tic@scilly.gov.uk

Ordnance Survey Explorer Map number 101
© Crown Copyright 2008

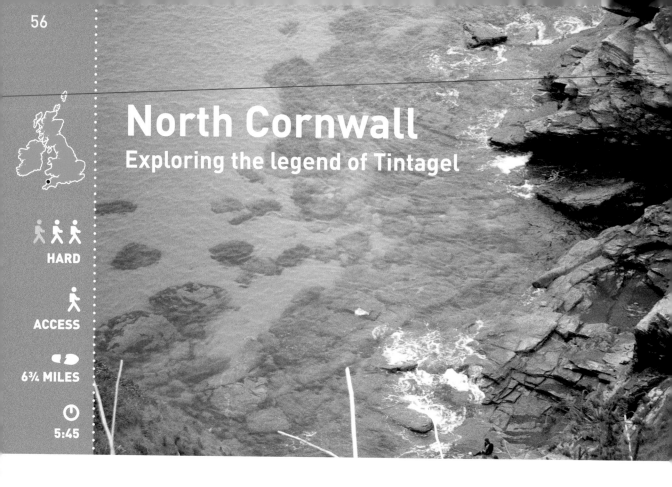

North Cornwall
Exploring the legend of Tintagel

HARD

ACCESS

6¾ MILES

5:45

This exhilarating walk covers a large section of the Cornish coastal path and presents the walker with spectacular rugged geology and abundant flora.

1 Start the walk by taking the narrow lane opposite the car park signposted to the church. On the way note the herringbone style 'stone hedges', locally called 'curzyway'. These hedges are a unique feature to this area. Throughout this walk you see abundant flora; sea and pink campion, common valerian, common toadflax, rosebay willowherb, sheep's bit, bramble, gorse, foxglove and smooth hawk's-beard. As you round the final bend in the road Tintagel Parish Church is to your right, directly fronting the coast. This Norman church was built between 1080 and 1150 and is dedicated to St Materiana (a princess from

Gwent) who evangelised in this region in 500 AD. The churchyard is entered via a Cornish 'stile' comprising stone cattle grid and coffin rest in the centre. In the graveyard there are several examples of quite distinct headstones, including a wooden cross and coffin-shaped gravestone. To the left of the church is the coastal path to the youth hostel.

② Turn right out of the churchyard and continue towards Tintagel Island and the castle. Take care when viewing the abandoned cliff slate workings not to get too close to the destabilised cliff-edges. Adjacent to this section of path there are easily accessible rock outcrops, which are described as Upper Delabole slate overlain by laminated black slate, both of Upper Devonian in age (about 370 million years old). These rocks and those of the island have been disturbed by faulting and, consequently, the time relationship between them and the tuffs and lavas of the Tintagel Volcanic Formation are also disturbed. As you continue, you will notice signs indicating level of difficulty as you proceed towards the causeway to **Tintagel Island** and the castle immediately ahead of you. If you have difficulty with heights it is best to avoid the descent to the island, which is near vertical in places.

③ If you decide to enter the castle and island there is a charge, but it is well worth it. The steps down to the island are extremely steep and slippery when wet. Follow the designated paths, noting the ruins on the way. When you reach the top of the island you will be rewarded, on clear days, with an uninterrupted panoramic view of the open ocean and headlands to the west.

Either after or instead of crossing over to the island and castle, head down to **Tintagel Haven**, which once provided a sheltered anchorage for the export of slate and imports of coal and everyday goods. The redundant whim above the beach was part of the winch used to haul slate to the ships waiting below.

These ships would be moored beneath the cliffs and slate would be lowered directly onto the vessels. This practice ceased in 1865 and instead vessels were beached at low water as this was considered to be less hazardous.

④ Follow the path above the beach by crossing the boardwalk above the stream that feeds a small waterfall. Looking back you will see two caves on the left flank of the beach. The smaller one was excavated for copper and lead in 1852 and, although worked for several years, failed to be profitable. The larger one is known as Merlin's Cave and is one of numerous references made to an Arthurian connection with the island and castle. Bordering the path heading down to the beach there is an outcropping light green coarse-grained slate and a warning of potential rock fall. Slightly inland from here is a small cafe and toilets, and a return route to Tintagel.

Ascend the steps and path out of the Haven and head towards **Barras Nose**. You can walk to the furthest point of this promontory but with care. The geology of Barras Nose is a mixture of cleaved rock types: slate and siltstones with volcanic lava and tuff at its furthest point.

⑤ From Barras Nose take the path that leads up towards Willapark (not to be confused with the Willapark to the east at Boscastle). Bypass the connecting footpath back to the town, via the Camelot Castle Hotel. As you proceed towards Willapark (Cornish for 'enclosure with a view') you pass Barras Cove and Gullastem below. Continue through a kissing gate and up Smith's Cliff

DELABOLE SLATE

There were several quarries in the vicinity of Tintagel that extracted this group of slates. Delabole Quarry is the deepest slate quarry in England and continues to operate today. The thin slatey cleavage lent itself to lightweight roofing and became commercially important by the 14th century, and probably earlier, for local use. By the early 17th century the north coast became the predominant Delabole Slate supplier both at home and abroad.

TINTAGEL CASTLE

Despite claims of an Arthurian connection, Tintagel Castle is mostly of 11th and 12th century Norman origin with some 6th-century buildings (the 'northern ruins'). In Cornish 'Tintagel' (*Tyn-tagell*) refers to a fortress of the narrow entrance (*Tyn-* meaning a fortifiable place and *tagel* meaning construction). The ruined monastery may date from 500 AD, which is contemporaneous with the evangelising mission of St Madryn in this area.

THE DIPPER

Populations of this small bird are relatively stable in the UK. There have been reports, however, of regional declines and more rarely local extinctions. The acidification of rivers by air pollutants and an increase in coniferous tree planting on riverbanks are given as the primary causes for their decline. The acidification means acidic leachate enters streams and rivers, and decimates the aquatic invertebrates upon which the dipper feeds. In the south-west they remain an uncommon sight, but are not endangered.

towards Willapark and its ancient settlement. You may be able to discern some signs of the Iron Age settlement, but most have been levelled and otherwise disturbed. Just offshore are **Lye Rock and** further west **the Little Sisters**. It was on the Lye Rock that the *Iona*, carrying coal from Wales to Trinidad, was wrecked on 20 December 1893. Four local men saved nine sailors, but three others drowned, including the cabin boy Domenico Catanese, who is buried at St Materiana.

6 Continuing east you follow the lower path around a small headland above Bossiney Haven. The steps leading down (and out) to the Haven valley are uneven and can be slippery when wet. The beach below is popular for kayaking and for exploring the caves below, but is only safe for swimming on the incoming tide. Continuing on, cross a small wooden bridge over a stream. Here there is a four-way marker showing the path to the beach, a return path to the main road to Tintagel via Bossiney village and our forward path. Take the 60 or so steps out of the narrow valley to reach Bossiney Cliffs above Benoath Cove. When at the far side above the valley look back and down to see the aptly named Elephant Rock. Continue east over Bossiney Common, a welcome level walk to **Rocky Valley**. On your approach to Rocky Valley you

will pass several large outcrops of cleaved and faulted greenish-grey Upper Delabole Slate that border the path. Some discrete areas are colonised by lichen and include large quartz nodes. The cliffs below show similar rock strata with distinct quartz veining clearly visible despite the distant view.

7 Continuing on you arrive at a seat above Rocky Valley, which provides a well-earned rest before the steep descent. This is a narrow V-shaped gorge with a discrete opening out to sea. The steps down are uneven and 'rustic' in construction. There is a wealth of flora bordering the valley sides that includes fine cliff grassland. Near the bottom the path levels out to a shallow gradient and borders the stream which tumbles its way down the rocky floor and over distinct changes in gradient. This small meandering **stream** has eroded its path through the confining rock as it travels out to sea. Off to the left of the main path another narrower one winds its way towards the sea opening. On this path are warning notices not to proceed seaward because of the potential hazard of being swept out to sea by waves on the incoming tide. Further along the stream you may be fortunate to observe dippers hopping from rock to rock and feeding their young. A short detour up the valley brings you to the remnants of Trevillet Mill and on the east side cut into the rock face are two 'maze' carvings, which have evoked a multitude of interpretations as to their origins. This east side is privately owned. A zigzag path leads you out of the valley.

8 Continue on, passing above Trewethet Gut and Trambley Cove towards Trevalga Cliff. As you proceed **Saddle Rocks and Darvis's Point** come into view. On passing over Trevalga Cliff bear right and past Firebeacon Hill. In the far distance and ahead you will see a white building standing on a promontory. This is the old lookout station above Boscastle. At this point go over a stile, head downhill and bear right. Proceed through a gate and up a track that leads to Trevalga village, passing the manor house to your right. The manor and village are now held in trust in order to protect them

from development. If time allows take a short detour to the Norman church of St Petroc's. At the crossroads turn right for Tintagel. This is a busy road and footpaths are rare, so care needs to be taken.

North Cornwall

ADVICE

This walk is not suitable for wheelchair users and buggies. There are a number of steep, uneven steps that are slippery when wet. Some of the inclines are also very steep and this applies especially to Tintagel Castle and Island. There are few barriers or notices en route to warn walkers of cliff falls/unstable cliff edges and great care, therefore, needs to be taken with each of these potential hazards. A hard hat should be worn if walkers intend approaching cliff faces and overhanging rock outcrops. Dress appropriately: this section of coast is, for the most part, directly exposed to the weather systems originating in the North Atlantic, particularly westerly gales.

Ordnance Survey Explorer Map number 111
© Crown Copyright 2008

PARKING

There are several pay-and-display car parks in Tintagel. However, if you want to avoid the road walk back by using the bus, it may be advisable to do the walk in reverse (starting at Trevalga village). Buses can be hailed anywhere along the B6263 route. The bus stops at Trevalga village and outside the Visitor Centre in Tintagel.

START

The walk starts from the car park near the old post office in Tintagel village.

CONTACT DETAILS

Tintagel Visitor Centre, Bossiney Road, Tintagel, Cornwall PL34 0AJ
t: 01840 779084/250010
f: 01840 250010
e: tintagelvc@ukonline.com
w: visitboscastleandtintagel.com

Lynton and Lynmouth
and the Valley of the Rocks

MEDIUM

ACCESS

6½ MILES

4:30　①

VALLEY OF THE ROCKS

The rocks in this 'dry valley' were probably created by frost weathering during the ice age, and possess some extraordinary formations. Dominating the seaward end is the tor-like Castle Rock, with Rugged Jack to the right of the valley and the prominent pinnacle of the Devil's Cheesering on the left. It is believed that this valley may have been the original exit of the Lyn rivers.

This walk uses recognised footpaths and provides a panoramic view of the Valley of the Rocks and some of England's most spectacular coastline.

① From the car park cross the road near the sheep grid to the seated viewpoint opposite and then follow the zigzag path up the hill (signposted to Lynton via Hollerday Hill) to reach a broad path. At this point look back for a wonderful panoramic view of the **Valley of the Rocks**. To your immediate right is Rugged Jack and, dominating the valley beyond that, Castle Rock. A herd of a rare breed of wild goats, brought to Britain over 6000 years ago, is often seen grazing in this area and Exmoor ponies roam across Hollerday Hill. Turn right along the broad path, following it towards Lynton. Continue slightly uphill, through a gate and woodland (mainly sycamore, blackthorn and some very large gorse bushes, with an under flora of geranium, rosebay willow herb and, in the spring, primroses). The path continues gently downwards through some mature woodland with a carpet of ferns, turning right at the signpost for Lynton.

② Just before passing through a rock cutting, and to the right, are the barely recognisable remains of Hollerday House. This mansion was built in 1893 by Sir George Newman, a major benefactor to Lynton, but, after his death in 1910, death duties resulted

in a decline in the property and it finally mysteriously burnt down in 1913. After the cutting the path runs onto a lane, which takes you down to the left-hand side of the **Town Hall**. Lynton and Lynmouth, which lie some 500 feet below, and the surrounding countryside, was referred to by Victorians as 'Little Switzerland', because of the profusion of deeply cut wooded valleys, rivers, waterfalls and high cliffs.

3 Turn left at the Town Hall and walk down Lee Road, past the Valley of the Rocks Hotel, to turn left before the parish church. From here proceed downhill, following the signpost for Lynmouth. As you descend, you have a fine view of the bay and the mouth of the River Lyn. At the junction of paths, choose the one signposted B3234, passing Bay View House to reach the road. Turn left, walking down the right-hand side of the road to cross the bridge over the West Lyn

River and past the entrance to the Glen Lyn Gorge (this has a riverside walk to waterfalls and the point where particularly severe devastation occurred during the 1952 floods). Turn left to cross the stone bridge over the **East Lyn River** and then immediately right into Tors Road.

4 Walk along Tors Road until you reach the river path (signposted to Watersmeet via Riverside Walk), which climbs uphill above the river, before descending to Blackpool Bridge. Cross this, turning left to walk along the right-hand bank of the river (look out for wagtails on the rocks in the river). Go past the first bridge (Lynrock Bridge) on the opposite bank of which, where a spring issues from a rock, is the remains of the Lynrock Mineral Water Works. This was destroyed in the 1952 floods, as were many of the bridges, which were used as packhorse bridges to convey wood products and lime along this, one of Britain's deepest gorges. The gorge was eroded by the waters of the East Lyn River over many centuries as sea levels fell when the ice sheets were created. Just after Murtleberry House join a tarmac lane, which takes you to the side of a rustic stone bridge. Do not cross, but carry straight on, eventually bearing left to cross two wooden bridges. The first of these crosses the Hoar Oak River, upstream of which are a number of **waterfalls**. The second bridge crosses the East Lyn River to take you to Watersmeet House on the other bank. This is an old Victorian fishing lodge, built around 1830, now converted into an attractive teashop and garden operated by the National Trust.

5 Proceed, with the house on your left, past the toilet area and take the path into the woods (signposted to Fisherman's Path). After about 150 yards be careful to look out for a few steps going up left onto a path, which rises backwards into the upper wood. Follow the path, which rises steeply in a zigzag, up through the oak woodland. At a path junction, take the left, more recognisable, path, eventually arriving at a clearing at the top of the climb. In front of you is the slope

LYNMOUTH FLOOD

Devastating flooding occurred on 16 August 1952 when, after 24 hours of torrential rain over Exmoor, the East and West Lyn Rivers came together as a torrent. The riverbanks were breached, causing cottages, a chapel and shops to be swept away. Thirty-four people died and Lynmouth had to be evacuated for some time, with hundreds left homeless.

VICTORIAN CLIFF RAILWAY

In early Victorian times it was extremely difficult to move between Lynmouth and Lynton. Freight and people had to be moved by packhorse and carriages. The original concept of the railway was developed in the 1880s and, finally constructed by the engineer George Marks, opened in 1890. Its unique features were four separate breaking systems which could stop either car independently and a hydraulic lift which, when the topmost car was filled from a reservoir, caused that car to move down, simultaneously lifting the other car upwards.

of Countisbury Hill. Head towards the hill until you reach a four-way signpost. Proceed through the gate ahead of you towards Countisbury. Climb the open hillside, past the waymark post, and through the wooden gate of the sheep pen in the far right-hand corner of the field. Head downhill, emerging shortly onto the A39, turning left here to walk a few yards as far as the Exmoor Sandpiper Inn.

In 1899, with a vessel in distress off Porlock Wier, and because of impossibly high tides, the citizens of Lynmouth decided to haul their lifeboat 13 miles to save the threatened sailors. This inn was where many lost heart, after hauling the lifeboat on a massive horse-drawn carriage up the hill from Lynmouth. However, some resilient souls heroically carried on across the moors to Porlock, to launch the boat the next morning and save the crew of the foundering vessel.

6 Cross the road and walk towards the **church**, passing through the graveyard to emerge at the rear through the wooden gate. Turn immediately left to join the coastal path, which takes you around a wall and downwards across the hillside. After about 200 yards turn right down some steps and left around the stone wall, eventually passing through a wooden gate. Stop here and look back along the shore to Foreland Point, which contains some of the oldest rocks in Devon. The red-coloured Devonian sandstone, making up much of this coastline, was laid

down 400 million years ago. Below you are the Lynton Beds, a collection of mudstones, siltstones and sandstones, and where they meet the Foreland Grits, the geological fault is known as the Great Red. Continue along the path with a fine view of Lynmouth and the coast beyond, and near a gate and seat you are walking below the old linear defensive earth ramparts of Countisbury Castle, which sat on the hill above you. As you approach the woodland, it is necessary to emerge onto the road, following the narrow, clearly waymarked path along the roadside for a short distance, before heading down into the wood. Keep following the coastal path to eventually turn right towards the beach.

7 When you emerge into the lane, turn right and then left, following the path signposted to Combe Martin, along the edge of the beach. Just after the putting green, bear left and then right, to cross the footbridge over the harbour. Turn right past the public toilets and walk a short way around the harbour side to the **Cliff Railway**.

8 Ascend to Lynton, either by the railway or by the steps and path just before the railway entrance. The continuously steep tarmac path crosses the railway a number of times, giving good views of the trains. Emerge from the pathway (near the parish church) or from the railway (on the uphill side of the Valley of the Rocks Hotel) into Lee Road and turn right to retrace your steps to the Town

Hall. Turn right immediately before you get there, retracing your route up the lane and pathway towards Hollerday Hill. After about 300 yards, where the main track bends left, carry straight on along the coastal path, through a kissing gate, to walk along the seaward side of the hill. There are fine views from here to the Welsh coast, the panorama stretching from the Glamorgan Heritage Coast opposite to the Gower peninsula in the west. When reaching the point overlooking the Valley of the Rocks, turn right and descend the zigzag path to your starting point at the car park.

Lynton and Lynmouth

ADVICE

Because of the inclines (there are three steep climbs, one in the Valley of the Rocks, one through the woods after Watersmeet and one from Lynmouth to Lynton, if you don't use the Cliff Railway) this walk is not suitable for wheelchairs or baby buggies. Good walking footwear is essential as many paths are uneven.

PARKING

There is a car park in the National Park Authority's picnic area on the left-hand side of the road, just before the cricket field. Toilets are available here. There are also buses: (No.300) from Taunton to Lynton and a First Bus (No.309) from Barnstaple to Lynmouth.

START

Start at the Valley of the Rocks, ½ mile out of Lynton.

CONTACT DETAILS

Lynton Tourist Information Centre, Town Hall, Lee Road, Lynton, Devon EX35 6BT

t: 0845 6603232

f: 01598 752755

w: lynton-lynmouth-tourism.co.uk

Ordnance Survey Explorer Map number OL9
© Crown Copyright 2008

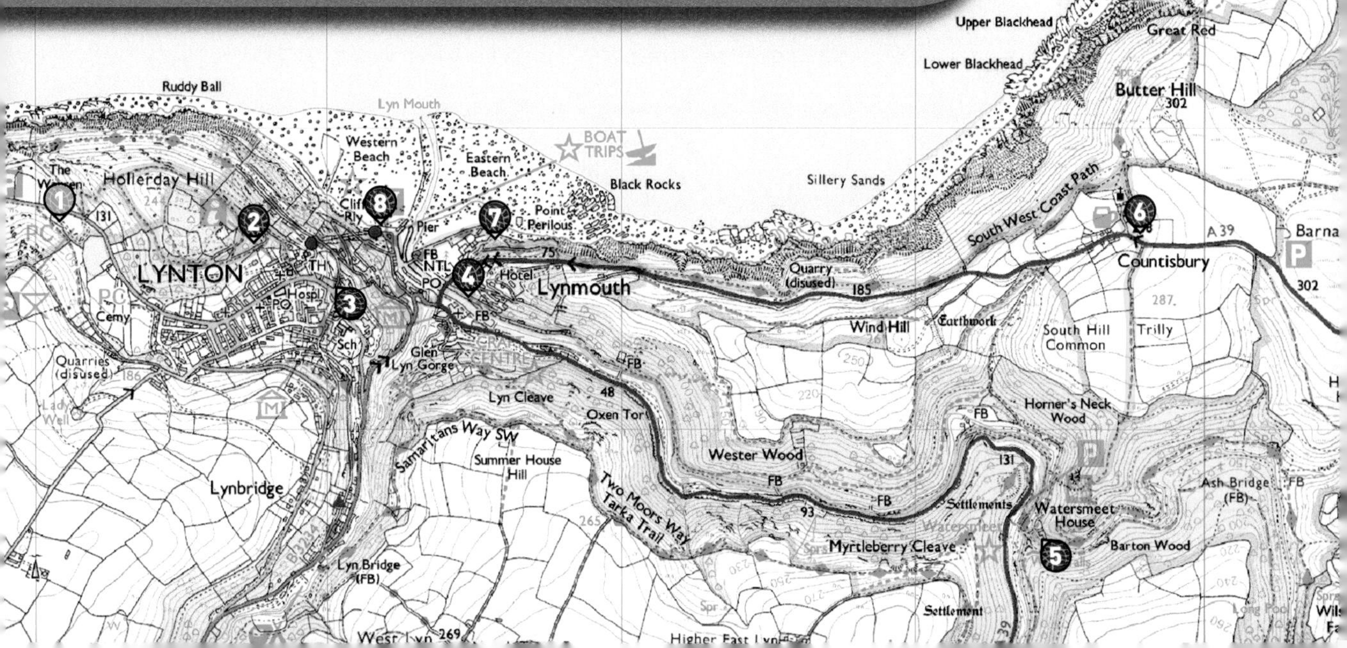

Bristol Docks
A monument to Brunel

EASY

ACCESS

4 MILES

3:00

Isambard Kingdom Brunel, one of the 18th century's most versatile and inspired engineers, played an important role in the development of the Bristol Docks. This walk showcases some of his finest work.

Bristol's name arises from a bridge that crossed the river Avon some thousand years ago in the vicinity of the present Bristol Bridge. That original bridge was called *Brig-Stow*, meaning ' the place of the bridge'. In the middle of the 12th century, an abbey of canons regular, dedicated to St Augustine, was founded by Robert Fitzharding. His abbey later became Bristol Cathedral and the surrounding area, extending to the waterfront, has been known for centuries as Canons Marsh. It is around this area that Bristol grew and the docks developed.

❶ Start the walk at Temple Meads station. Leave the present station, designed by Brunel's colleague, Sir Matthew Digby Wyatt and completed in 1878, and walk down the right-hand side of the station yard. Brunel's original station (now the **British and Commonwealth Museum**) is on your right. The three-storey entrance, in Temple Gate, gives it the appearance of a Tudor mansion. Completed in 1841, it is believed to have been the world's first true railway terminus, housing everything below one great arched roof. At the bottom of the slope turn right and immediately left over the pedestrian crossing to cross Temple Gate towards the Holiday Inn. Turn right and follow Temple Gate around to the left into Redcliffe Way, following the sign for St Mary Redcliffe.

② Before the church, and on the opposite side of the road in front of the Open University, is Chatterton House. Here the boy poet, Thomas Chatterton, was born in 1752. With his talents unrecognised in Bristol, he left home to die tragically in London in 1770, giving rise to the legend of a 'boy genius' destroyed by a philistine world. **St Mary Redcliffe**, to your left, is one of the largest parish churches in Britain. Its spire, destroyed by lightning in 1442 and rebuilt in 1872, is the second highest in the country. It survived the violent bombing of the Second World War, being something of an icon to Bristolians at that time. At the roundabout, bear slightly left and then right across Redcliffe Hill, via the pedestrian crossing. Turn right after crossing and continue around to the left, following the sign for Arnolfini, to cross over the bascule Redcliffe Bridge. This bridge crosses the upper reaches of the floating harbour, whose route you will follow as far as the River Avon. At the end of the bridge turn left along the Grove and left again after the Riverstation Restaurant to follow the dockside as far as the Prince Street Bridge.

③ Just opposite is the Arnolfini gallery, created in 1961 to promote interest in

contemporary arts, and, on the far corner of the quayside, is a statue of John Cabot. He sailed from here in 1497 on the *Matthew* to search for the Far East, but, sailing westwards, discovered Newfoundland. Cross the Prince Street Bridge and turn right past the Bristol Industrial Museum and the tall electric cranes, to follow the Bristol Harbour Railway to eventually arrive at the Maritime Museum and SS *Great Britain*.

④ To reach there, you walk along St Augustine's Reach, which was created in 1247 when the River Frome was diverted. In 1810 the channel was enlarged and the Avon diverted, with lock gates at the western end isolating the floating harbour from the ebb and flow of the huge tides. Trading from this area of the docks increased considerably between the 16th and the 18th centuries, when the docksides would have been crowded with ships, many of which would have followed the triangular trading route between Europe, Africa and America. European goods were exchanged for people in West Africa, who were transported as slaves to the Caribbean, from where sugar, cotton and tobacco were brought back to Britain (the huge **tobacco factories of W.O. Wills** can be

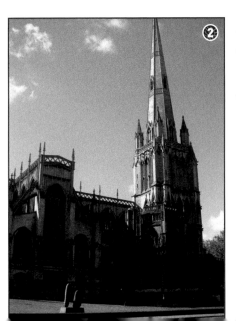

SS GREAT BRITAIN

Launched in 1843, she was the first screw-propelled, wrought iron ship, although Brunel originally conceived her as a paddle steamer. She ran aground in Ireland in 1846 but, after rebuilding, spent 24 years as a fast passenger ship. In 1855 the ship was used as a troopship in the Crimean War, but then, by 1886, she was downgraded to a sailing ship and coal carrier, eventually becoming beached in the Falklands. There she remained until salvaged and towed to Bristol in 1970, to be restored by 2005 as a monument to British invention.

BRISTOL AND THE SLAVE TRADE

From the 12th century Bristol had been involved in transporting children to sell as slaves in Ireland. Bristol's merchants, already rich in the Middle Ages, began illegally trading in slaves from Africa to the Caribbean in 1670. The trade was made legal and was conducted on a large scale during the 18th century. By 1730 Bristol was Britain's main slaving port, but when the Abolition movement gathered momentum many local Bristolians became involved and the trade had ceased by 1807.

BRUNEL'S

SLUICES

By 1830 the floating harbour was suffering from severe silting. Brunel constructed three shallow sluices and one deep scouring sluice between the harbour and the Avon's New Cut, and commissioned a dredging vessel. At low tide in the Avon, silt was sucked through the sluices to be carried away by the tide. The sluices were rebuilt in the 1880s and are used today. They are housed just beyond Underfall Yard.

seen a little further along). Shipbuilding also took place along this basin and, just alongside the Industrial Museum, where you have walked, Brunel's oak-hulled paddle steamer, the SS *Great Western*, was launched in 1837.

5 Fifty yards further on you will come to the reconstruction of John Cabot's *Matthew* (though it is sometimes moored near the Redcliffe Bridge). Continue the walk between the Maritime Museum and the SS *Great Britain* up Gasferry Road, turning right at the corner of a derelict Victorian malthouse (and virtually opposite Caledonian Road) along a path. Follow this back towards the waterfront, turning left at the public toilets and then right onto the path around the marina, eventually turning left along the waterfront.

Continue along the harbour, passing the Cottage Inn, to Underfall Yard. Walk through the yard (where boat maintenance still occurs), turn left out of the gate and right at the road. At the end of the road cross to the car park opposite, go to its end, and turn left along the waterfront footpath, near a large anchor. You are now walking along the edge of Cumberland Basin, which was created when the engineer, William Jessop, built a dam and lock system to isolate the harbour from the Avon tides. Walk under the left-hand arch of the Plimsoll Bridge, around the spiral stairway and onto a grey tubular girder bridge across Brunel's now disused South Entrance Lock. Cross to the island and walk to the furthest point, at the edge of the Avon, for a view of the **Clifton Suspension Bridge**.

6 Cross the wooden lock bridge near you and turn right to walk along the edge of the 1873 Howard Lock, under the grey bridge (look across to see Brunel's original swing bridge preserved on the island quay). After about 100 yards, and before the next bridge, leave the quayside at Cumberland Basin Road, crossing into the Pump House ahead of you, and then proceed along the waterfront to cross the metal footbridge over Pooles Wharf. Continue along the waterfront until you reach the ferry landing. At this point you have an excellent view across to the SS *Great Britain* and, if present, the *Matthew*. In 2007 the dockside path was closed just after this point so it is necessary to leave the quayside and proceed along Anchor Road. Walk to the rear of some old dock buildings and turn right down Canons Way and then, shortly, right again into Cathedral Walk to rejoin the dockside path on Hanover Quay. Turn left and follow the harbour past the Amphitheatre, the open area of which is used for shore-side entertainment events, and reach **Pero's Bridge** to cross the River Frome at Bordeaux Quay (indicating Bristol's importance in the wine trade). Pero's Bridge is dedicated to the memory of a Caribbean slave, Pero Jones, brought to Bristol in 1783 from the plantation of Bristol merchant, William Pinney. He became a servant of the Pinney family, who lived in what is now the Georgian House Museum in Great George Street.

7 Before crossing the bridge it is possible to make a short detour to the left, across Anchor Square to the right-hand corner, to a point opposite some steps (unsuitable for wheelchairs). Ascend these to visit **Bristol Cathedral**, which you can see in front of you. The cathedral, begun in 1542

and completed in its present form by 1888, is one of the finest examples of a Hall Church, the nave, choir and aisles all having the same height. Retrace your steps to Pero's Bridge, cross and carry on straight along Farrs Lane and Royal Oak Avenue into Queen Square.

8 **Queen Square**, named after Queen Anne when she visited Bristol in 1702, was home to many of Bristol's wealthy trading merchants, the statue in the middle being that of King William III. The Bristol Riots occurred here in 1831 and it was said that Brunel, visiting Bristol to supervise work on his suspension bridge, was sworn in as a special constable to help prevent looting in the square. Take the right-hand side through the square and emerge to cross Redcliffe Bridge. From here retrace your steps, past St Mary Redcliffe, to Temple Meads station.

Bristol Docks

ADVICE

This circular walk follows city pavements and pathways and, apart from some cobbled areas and the slight incline near Temple Meads station, is perfectly flat. Abundant cafes, restaurants and public houses occur along the whole walk, with public toilets being available at intervals.

PARKING

Easy access is available via Temple Meads station. Parking is available at a multi-storey car park opposite the access road to the station in Temple Gate or in the station car park.

START

Start in the car park outside Temple Meads station.

CONTACT DETAILS

Bristol Tourist Information Centre, Harbourside, Explore At Bristol, Anchor Road, Harbourside, Bristol BS1 5DB
t: **0906 711 2191**
f: **0117 9157340**
e: **ticharbourside@destination bristol.co.uk**
w: **visitbristol.co.uk**

Ordnance Survey Explorer Map number 155
© Crown Copyright 2008

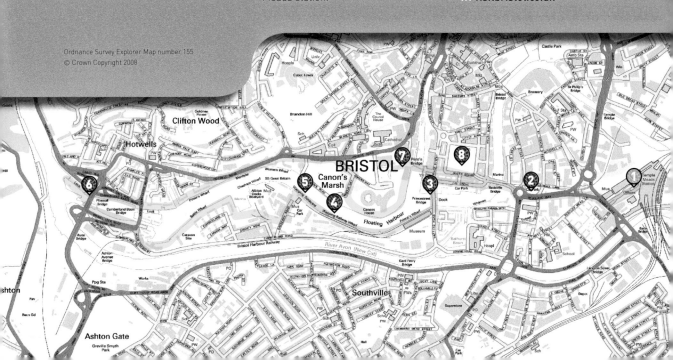

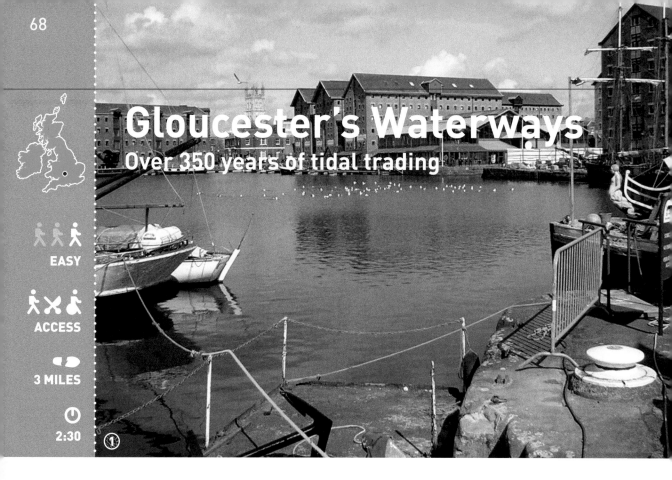

Gloucester's Waterways
Over 350 years of tidal trading

EASY

ACCESS

3 MILES

2:30

①

WESTGATE BRIDGE

This painting, by I. Harris, shows the original five-arch Westgate Bridge in 1806, three years before it was demolished.

You might not think of Gloucester as coastal, but the tidal waters of the River Severn flow here and the famous docks, once bustling with commercial shipping, are now busy with pleasure boats.

① Making your way through the old warehouses it is possible to imagine the bustling scene a hundred years ago. **Gloucester docks'** main basin is the terminus of the 16-mile long Gloucester to Sharpness Canal. The canal, opened in 1827 as the widest and deepest in England, was built to allow ships to bypass a difficult winding stretch of the River Severn, allowing Gloucester to become an inland port.

The old warehouses were built between 1827 and 1873 to store corn. Today the buildings are used as offices, housing, shops and the National Waterways Museum.

They have also been used for various films and TV dramas over the years, one of the most famous being *The Onedin Line*, which was filmed here in the 1970s.

② Make your way anti-clockwise around the main basin, heading towards **Gloucester lock**. This is the gateway between the northern end of the canal and the River Severn. Once busy with cargo barges, it is now a popular visitor attraction. The lock is 180 feet long and can hold up to six narrowboats. It takes about 15 minutes to lower the boats the 16 feet to river level.

Continue walking to the end of the lock, crossing the narrow road and keeping the river on your left. Stop where the river starts to gently bend to the left. You are now be at Gloucester quay.

3 The stone wall between the river and the road here marks the site of the **original riverside quay**, which dates back to ancient times and is where cargo boats moored to load and unload their wares on their way to and from places like Bristol, Tewkesbury, Worcester, Bewdley and beyond. As far back as 1580 a Royal charter from Queen Elizabeth had granted Gloucester port status. The local authorities hoped to benefit from this by collecting dues on goods handled at the quay, but in practice only a few foreign-bound ships were seen at the quay because of the difficulties of navigating the shallow tidal stretch of the river approaching the city. It was not until the canal opened that Gloucester really began to develop as a port.

Continue along the footpath, following the road round to the left. Take care here – it is often busy with traffic. Carry on past the garage on the left and head straight on, following the blue signs towards Maisemore and Highnam, and cross over the river using the footbridge.

4 The modern **Westgate Bridge** spans the river at an important crossing point and is the real reason for Gloucester's existence.

The Romans settled and set up camp here at Glevum, as it was known, to control the river crossing to Wales, as this was the lowest point on the River Severn where a crossing was possible. It seems likely that this was also chosen as the site for the river's first bridge because it splits here into two narrower channels, forming an island (Alney Island), and it was thought that two short bridges were easier to build than one long one. The original medieval bridge had five great arches. This was replaced in 1816 by a single span bridge designed by Robert Smirke. Then in the 1970s two new wide-span bridges were built, while a new set of three bridges was completed in 2000.

NARROWBOATS AND BARGES

Many people mistakenly call the long, thin narrowboats (popular today as pleasure boats) barges. In fact barges were wider, allowing them to carry more cargo. Narrowboats are only 7 feet wide enabling them to use the narrow locks of the Midlands' canals.

Cross the bridge and follow the blue signs, towards Maisemore, using the footpath along the side of the main road. Take care as the road is often busy. Carry on for about half a mile and follow the footpath as the road curves to the left (blue signs to Ross/Chepstow). Cross the river on the modern road bridge. When you reach a gate on the left (signposted Gloucestershire Way) go through it, doubling-back onto the old stone-built bridge.

5 From this disused bridge you will get a bird's-eye view of Gloucester and the surrounding countryside. **Over Bridge**, designed by the Scottish civil engineer Thomas Telford, was completed in 1829 and replaced a 16th-century eight-arch bridge that had been damaged by ice in 1818. It's now preserved as the oldest large-span masonry road bridge in England.

This point is approximately halfway round the route. At this stage wheelchair users or buggy pushers should retrace their route back along the main road to the large road junction near Westgate Bridge. Here turn right and follow the footpath along the road for about half a mile, then turn right just before you get to the new road bridge along the cycle path, near the end of the walk.

To carry on with the main route, continue over Telford's bridge then down the steep embankment which descends to river level. Turn left at the bottom and head back under the bridge. Follow the path south along the river, taking care on the uneven surface.

This corner of Alney Island, known as Lower Parting, is where the two channels of the River Severn join together again. It's a great place to watch the sea birds that gather on the sand banks and mudflats, which are visible at low tide.

6 It's also a great place to view the **Severn Bore** – one of the natural wonders of the UK. The Bore is a large tidal wave that works its way up the Severn Estuary from Awre to Gloucester twice a day on 130 days of the year. At up to 6 feet high it cascades through 25 miles of countryside at speeds of up to 13mph.

From here you can see the wave coming towards you head-on before splitting off to either side. But if you're planning on doing this walk at a time when the Bore is due PLEASE BE CAREFUL. If the river is full, most likely in the spring or autumn, the water could overflow onto the point where you are standing.

Each year dozens of surfers compete to ride the Bore. Dave Lawson from Gloucester is the current champion, boarding for an official record of 5¾ miles.

7 Turn left and continue along the path, keeping the river on your right until you are able to see Gloucester Cathedral in the distance. Much of Alney Island is a **nature reserve**. The flood meadows here provide traditional wet grassland and marshy areas that are ideal for all sorts of wildlife.

Continue, keeping the river on your right. Just after you pass two pairs of telegraph poles cut across the meadow to join the new cycle track. Turn right along the cycle track, keeping to the left when the track splits in two. Just before the track goes under the new road bridge turn sharp right, almost doubling back on yourself along the grass track. Head towards the old railway bridge.

8 Make your way under the bridge, which goes over the east channel of the river, and continue a short distance until you find yourself between the walls of the old disused lock. This old lock – known as **Llanthony Lock** – was built in 1871, and was needed to allow boats passing up

and down the River Severn to avoid the new weir just south of here. The weir was constructed to ensure that there would always be a minimum depth of 6 feet of water in the river above Gloucester. It remained in use until 1924 when the incredibly high walls of the lock (required to cope with the flood level) began to move inwards towards each other.

Retrace your steps back to the cycle track. This is the point where any wheelchair users or people with buggies rejoin the main route. Continue under the new road bridge and follow the signs back to Gloucester Docks.

Gloucester's Waterways

ADVICE

Much of the walk is on tarmac, but part of it is on uneven grass, which can get muddy in bad weather, so an alternative is suggested for wheelchair users and buggies. It is mainly on flat ground with one short, steep descent on the main route.

PARKING

There is a pay-and-display car park at Gloucester Docks.

START

Gloucester Docks (signposted 'Historic Docks' from main road approaches to the city centre, and also from Gloucester railway and bus stations).

CONTACT DETAILS

Gloucester Docks Trading Company, Albion Cottages, The Docks, Gloucester GL1 2ER
t: 01452 311190
f: 01452 311899
e: info@glosdocks.co.uk
w: gloucesterdocks.me.uk

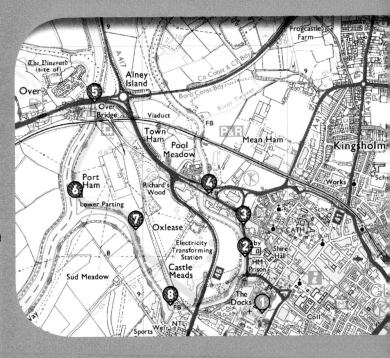

Ordnance Survey Explorer Map number 162
© Crown Copyright 2008

Cardiff Bay
From riches to rags and back again

EASY

ACCESS

1½ MILES

1:00

Once the largest coal ports in the world, Cardiff has experienced the highs and lows of industry, regeneration and immigration.

① The walk starts at the **Norwegian Church**, a distinctive wooden building known locally as the tube, because of its design. It was originally built in 1869 at West Bute Dock, about a mile from its present position. The church was a seamen's mission, but it was also a home-from-home for Scandinavian sailors, where they could read newspapers and magazines from home, write letters to their loved-ones, relax and chat with their friends. The family of writer Roald Dahl attended the church and he was baptised there in 1916.

As Cardiff Docks declined, so did the congregation. The church was finally deconsecrated in 1970, falling into disrepair through lack of maintenance and vandalism.

In the mid 1980s Dahl led efforts to set up a preservation trust to rescue the derelict building and to relocate it in the heart of Cardiff's historical docklands.

The church was dismantled in 1987 and moved here. Roald Dahl was the trust's first president, but he died on 23 November 1990, before the reconstruction was completed. The church was re-opened by Princess Martha Louise of Norway on 8 April 1992. Though it has not been consecrated, it is now used as a venue for concerts, exhibitions and wedding receptions.

2 As you leave the church make your way along the waterside where you will soon see the stunning new Senate Building, home of the National Assembly for Wales. Designed by Richard Rogers, it was opened on 1 March 2006 and is well worth a visit.

Just past the Senate Building you will see a magnificent terracotta building. **The Pierhead** was designed and built in 1897 by William Frame as the headquarters of the newly named Cardiff Railway Company and bears a coat of arms on its façade with the company's motto, 'wrth ddwr a than' (by fire and water), which celebrates the power that transformed Wales. The building became the administrative office for the Port of Cardiff in 1947.

Incorporating a French-Gothic Renaissance theme, the Pierhead boasts hexagonal chimneys, friezes, gargoyles, and a highly distinctive clock tower. Its exterior is finished in glazed terracotta blocks.

3 Go over the bridge to the left of the Pierhead building, towards Mermaid Quay, a recently redeveloped part of the waterfront now populated by bars and restaurants. Here sits **a bronze sculpture of a young couple** with their dog created by John Clinch in 1993. Entitled *People Like Us*, it celebrates the people who have lived and worked in Cardiff Bay.

During the 19th and early 20th centuries, Butetown, or Tiger Bay as it is also known, was a thriving dockland area and home to a hugely cosmopolitan community, with over 50 nationalities settling here. The kaleidoscope of settlers helped build the docks, work the ships and service this industrial and maritime city. Singer Shirley Bassey, one of the area's most famous inhabitants, is a good example of this multicultural mix as she had an English working-class mother, Eliza, and a Nigerian sailor father, Henry.

4 From Mermaid Quay walk straight on towards Stuart Street until you get to Woods Brasserie, the former **pilotage building**. This was the administrative centre for the river pilots who operated in the Bristol Channel.

The pilots were vital in helping ships negotiate the treacherous waters. They were employed for their local knowledge to guide the large ships into the dock through narrow and shallow straits. They knew where the rocks and sandbanks were, which side of the buoys to pass, and they were familiar with weather conditions and the rise and fall of the tide – the Bristol Channel has the second highest in the world.

The building's origins are steeped in mystery, its thick stone walls making it distinct from the other buildings close by. Local historian Professor Neil Sinclair suggests that it may have been used to stable the workhorses that pulled barges down the Glamorganshire Canal.

5 Turn left on passing the old pilotage building, then right just before you get to Harry Ramsden's, and you will come to the waterside at **Mountstuart dry docks**. During the First World War the docks were

ALL THAT JAZZ

Tiger Bay was a magnet for people who were really into their music and home to many clubs and pubs, such as the Casablanca, the Quebec, the Big Windsor, the Westgate, the Bute, and the Ship and Pilot. In fact, there were over a hundred pubs in the area, many of which have now disappeared. As well as the young Shirley Bassey, musicians who'd keep the locals entertained included guitarists Victor Parker, and Joe and Frank Deniz. The area was a haven for black American servicemen who were often shunned by communities near their bases in England.

TECHNIQUEST

Anyone who thinks science is boring should pay a visit to Techniquest to be proved wrong. This hands-on, interactive science centre is a major source of fascination for children and adults. There are more than 150 exhibits to play with, including giant puzzles, computers and a mirror maze. In addition, the centre has a planetarium, science theatre, cyber-library, a discovery room for young children, a shop and cafe.

extensively used for the maintenance of the British shipping fleet. Neil Sinclair sketches the scene during the dock's heyday: 'The whole area would have been full of ships. They would have to pick their time to dock very carefully. Ships could only sail in and out of the docks on the tide. Once the ships were in, the lock gates would close and when the tide went out the pumphouse at the entrance would drain the water from the dock.

'If a ship had severe engine problems it would be lifted out by crane and moved to the engineering shed. What is now the Techniquest building may look modern, but it is built on the foundations and walls of the giant repair shed. The authorities couldn't demolish it because it's listed. So they decided to build on top of the original structure, which is why it has such a unique shape.'

6 From here bear left and then turn right directly before the old pumphouse. Turning left, walk along Havannah Street, bearing right along a footpath when you are opposite the entrance of St David's Hotel, to a large bronze sculpture of a rope knot. From here you can carry on into the **wetlands**.

The building of the barrage in 1999 to create a freshwater lake to replace the mudflats and salt marshes was a key element in the plans to redevelop Cardiff Bay. The barrage serves to dam the mouth of the rivers Taff and Ely, resulting in a 500-acre freshwater lake and 8 miles of waterfront.

There was a lot of resistance to the plans from local householders who feared the barrage would raise the natural level of water

and result in floods. It took five separate Bills until Parliament finally passed the Act that would pave the way for the redevelopment.

The exclusion of seawater from the bay has changed the area's flora and fauna. Large waterfowl, wading birds and grey mullet have relocated to newly protected areas, including the nearby Gwent Levels Reserve just outside Newport, and have been replaced by freshwater species and a new wetland habitat.

7 Retrace your steps to the dry dock. When you get to Techniquest turn left through a gap in the wall onto Stuart Street. Cross the road, bear left and then turn right and walk down Adelaide Street, then cross James Street into Mountstuart Square.

Once a residential square, this area became the centre of the coal trade during the middle of the 19th century and was home to the **Coal Exchange**. Following its opening in 1886, coal owners, ship owners and their agents met daily on the floor of the trading hall. During the peak trading hour between midday and 1 p.m. the floor might have had as many as 200 men gesticulating and shouting. It was thought that up to 10,000 people would visit the building each day. At one time the price of the world's coal was determined here.

However, with Cardiff so overwhelmingly dependent on a single product the city was highly vulnerable to any downturn in demand, a fact painfully apparent in the inter-war years. With the end of the Second World War the docks went into further decline. The Coal Exchange closed in 1958 and coal exports came to an end in 1964.

8 From the Coal Exchange turn right into James Street and head towards the Wales Millennium Centre in Roald Dahl Plass.

Roald Dahl Plass, or the Oval Basin as

The last of Cardiff's five docks, the Queen Alexandra Dock, opened in 1907, by which time the city was the greatest coal exporting port in the world. Little more than 75 years later Cardiff Bay had become a neglected wasteland of derelict docks and mudflats.

As part of the area's regeneration, the Oval Basin was filled in to create a public space, which was renamed Roald Dahl Plass in honour of the Cardiff-born writer. But the most significant addition to the area has been the landmark arts venue, the **Wales Millennium Centre**, or the Armadillo as it is sometimes known, which opened in November 2004.

From the Millennium Centre carry on and you can make your way back to the start along Harbour Drive.

it was formerly known, was the seaward entrance to the West Bute Dock, once the biggest masonry dock in the world. The dock was opened on 8 October 1839. Further docks were built to serve the rapidly increasing iron and coal trade, including East Bute Dock (1855), Roath Basin (1874) and the Roath Dock (1887).

GRAND OPENING

The Oval Basin opening was a grand affair. An early morning parade started from the castle to the trumpet blasts of the Glamorganshire Band and the bells of St John's. Masons, labourers, gentlemen and tradesmen made their way to the dock. The arrival of the giant ship *Manalaus* from Quebec was proof that Cardiff had arrived.

Cardiff Bay

ADVICE

Most of the walk route is paved; some areas have wooden decking. There are stone stairs with adjacent ramps for wheelchairs.

PARKING

There is a pay-and-display car park next to the Norwegian Church Arts Centre off Harbour Drive. Alternative car parks are located at Mermaid Quay off Stuart Street and at St David's Hotel off Havannah Street.

START

Start at the Norwegian Church Arts Centre, which is located on Harbour Drive on the waterfront in Cardiff Bay.

CONTACT DETAILS

Cardiff Visitor Centre, The Old Library, The Hayes, Cardiff CF10 1WE
t: 0870 1211 258
e: visitor@cardiff.gov.uk
w: visitcardiff.com

Ordnance Survey Explorer Map number 151
© Crown Copyright 2008

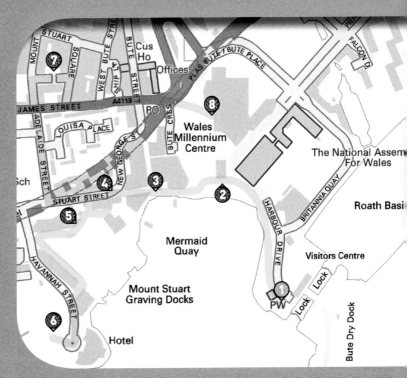

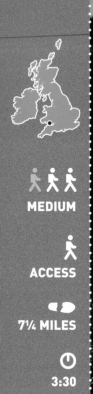

Glamorgan Heritage Coast

A geological journey

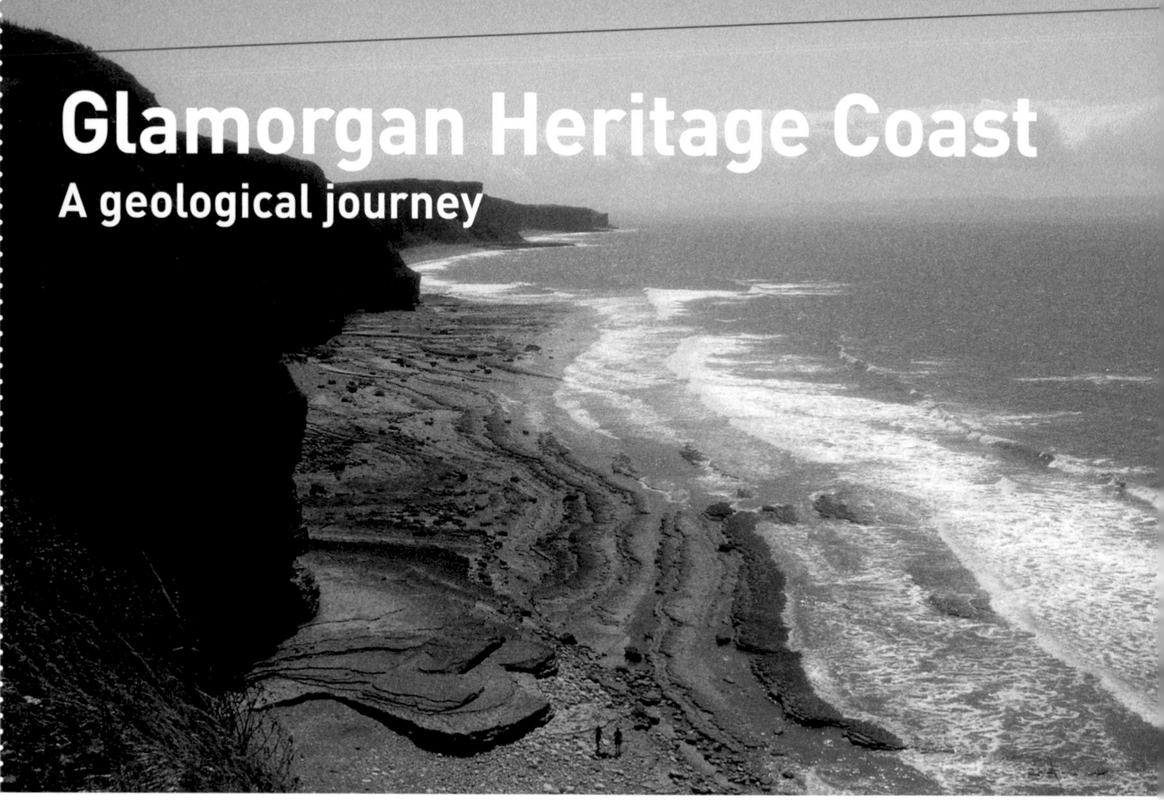

MEDIUM

ACCESS

7¼ MILES

3:30

FOSSILS

The most common fossils in the carboniferous limestone layer are the large, cylindrical *Siphonophyllum*. Other fossils to look out for are fan-shaped *Michelinia*, and brachiopods, which are spiral markings on rock surfaces caused by worms (*Zoophyces*).

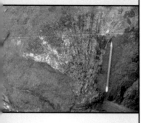

Much of this circular walk is along recognised footpaths, and includes a stretch of the Glamorgan Heritage Coast, woodland, agricultural land, sand dunes and a river estuary.

1

From the Rivermouth car park take the exit gate at the south-eastern corner and proceed along the coastal path. The Glamorgan Heritage Coast was the first stretch of coastline in Britain to be so designated and it has a fascinating changing geology as you walk along. Below the car park, and along the first part of the walk, are **beds of carboniferous limestone** which formed when this part of south Wales was in the southern hemisphere, near to where the Great Barrier Reef of Australia lies today, so corals are common fossils in these rocks. Also noticeable are Triassic breccias, which cut through the limestone, the most red of which can be seen just south of the car park. When these were laid down 200 million years ago, the land area would have been desert-like, lying just north of the equator. At that

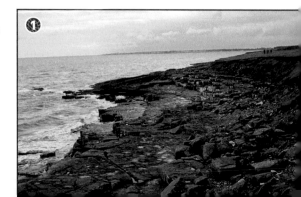

time desert wadis washed red sand, pebbles and rock inclusions over the carboniferous limestone and these were fused into the Triassic conglomerate seen overlying the rocks from the car park southwards for 300 yards. Look out for chuffs on the downs and herring gulls and oystercatchers on the rocks.

2 Just opposite the second stretch of wall, level with the large apartment block, are limestone platforms and the most prominent one is the best place for viewing fossils. Continue along the wide path after the wall, the rising gorse-covered slope to the left having been formed by overlapping Jurassic rocks 180 million years ago. These continue to form the cliffs you see in front of you stretching away to the south-east (note the layered nature of the Jurassic Lias cliffs). Just before the sign warning you about the dangerous cliffs follow the path up leftwards into the **gully**, which was formed by glacial run-off water. Exposed, on the right side of the gully is Sutton Stone (Littoral Lias), an unconformity of limestone conglomerate, laid down on the carboniferous platforms. This basal Jurassic exposure is one of the longest in Great Britain.

3 Before reaching the top of the gully, turn right onto the path that climbs the gully side to take you around a wall. Follow the path along the cliff-top, keeping the renovated farm building and walls to your left, heading eventually for the car park near the three seats. Enter the car park via one of the kissing gates and walk across to the narrow exit in the far right-hand wall. Pass through and after 30 yards cross the roadway to follow the path down to the beach, arriving eventually at the bottom of the steps near the car park in **Southerndown Bay** (also known as Seamouth).

4 From the steps turn right and immediately left past the lodge, following the tarmac track uphill to the walled garden. Enter the garden through the wooden door and follow the path around the three parts of the garden to emerge at the right-hand corner of the third garden. Proceed along a short woodland path, emerging at a gate, and turn right to reach the look out point. From here you have an excellent view of the Jurassic cliffs looking south-east along the Bristol Channel, with the Somerset coast and Exmoor to your right. Continue along the track to emerge at the top of the hill (Witches Point) at the ruins of Dunraven Castle. The view to the west takes in Porthcawl and, beyond, the **Gower peninsula**. Just below the ruin, on the northern side of the hill, is an Iron Age fort. Retrace your steps, following the tarmac track to the right of the ice tower (where Victorians stored winter ice for food preservation) and around the walled garden, back to the car park and the bottom of the steps.

Head past the Heritage Coast Centre to enter woodland by a stile, following the path uphill for about half a mile to cross another

DUNRAVEN CASTLE AND VICTORIAN GARDEN

The first castle on this site was built around 1128 and was destroyed by Owen Glyndwr in the 15th century. A manor house erected in the 1540s developed into a fine mansion by 1887. Used as a convalescent hospital in both World Wars, it was blown up by the Dunraven Estate in 1962 when no buyer could be found. The remaining walled garden was restored when the Heritage Coast became designated. The three gardens represent a Victorian garden (greenhouse), Victorian plant hunters and the story of fruit culture.

SAND DUNES

The Merthyrmawr complex shows the many phases of dune development – fore dunes, white dunes, grey dunes, dune grassland, scrub and woodland. Fore dunes are poorly represented, as the line of dunes stretching away from the rivermouth is subjected to extreme wave action. Well-established dune ridges run parallel to the river valley. These rolling sand hills are covered with red fescue and other grasses, with tall ribwort plantain rising from the moss. Common horsetail is present, with an extensive covering of dewberry and clumps of burnet rose.

stile and emerge into a field. Keeping the hedge on your left, follow the path to the top of the field, going left and immediately right over a stone stile. Cross another stile in the right-hand corner of this field and then follow the left-hand boundary of the further field. Climb yet another stile and continue along the left side of this fourth field to a stone stile. Turn right over the next wooden stile, cutting across, before the farm buildings, to a stone stile in the right-hand corner, emerging eventually at the road opposite the Farmer's Arms pub.

5 Turn left past **Pitcot Pool** (a favoured over-wintering area for wildfowl) and walk through the village of St Bride's Major. Just after the Fox and Hounds pub car park turn left towards the church.

6 The chancel of the Norman church of St Bridget was built in the 12th century by Simon de Londres, with the remainder being completed by the 14th century. It contains a number of monuments of various kinds installed in memory of past benefactors. St Bridget, the daughter of an Irish Chieftain, lived in the fifth century and founded several religious communities.

Continue up the hill (the Shilly) to the right of the church, shortly crossing a sheep grid. Immediately turn left along the drive leading to St Bride's Court, but before the gate take the stone stile in the left corner. Cross the field slightly uphill to the right-hand corner and cross another stile, turning left to continue through two fields. At the bottom left corner of the second field cross a stile and proceed along the left-hand side of the field to another stile

in the corner, which emerges to cross a narrow tarmac drive. Head for the waymark post at the outside corner of the field ahead and turn right, keeping the fence on your right (there are good views of the **Glamorgan hills** and valleys from this point). Head along the bridleway, towards a short stand of trees.

7 Bear right down the shallow valley, following the obvious sandy track towards a mixed woodland, bearing right and left just before it to reach the main road. You have been walking on the periphery of one of the major sand dune areas in Wales, which stretches some 20 miles along the coast of Swansea Bay. The slopes ahead of you across the river valley include some of the highest **sand dunes** in Europe. They were formed 10,000 years ago when melting ice transported Pennant sandstone from the coalfields. The dunes host some 321 species of flowering plants and ferns, about a quarter of the entire Welsh flora.

8 Cross the road and turn left, keeping to the obvious path to the right of a stone seat. The view back up the **river valley** is particularly pleasing. In the near distance is Ogmore Castle, which was built by William de Londres, one of the Norman conquerors of Glamorgan. It guards a ford, marked by stepping stones, and was one of a series of defences guarding the fertile vale of Glamorgan from attack from the sea. Continue following the path seawards and slightly upwards to eventually return to the Rivermouth car park. On route, keep a lookout for herons and cormorants along the river.

Glamorgan Heritage Coast

ADVICE

This walk is not suitable for wheelchairs or baby buggies, and care is needed along the section of cliff-top path if accompanied by younger children. Toilet facilities are available at the start of the walk and at Dunraven Bay. Refreshments may be available in Ogmore-by-Sea shops, at a small kiosk at Dunraven Bay (during the summer months) and at public houses in the area. Dogs should be kept on leads as sheep roam the common land areas.

PARKING

There is ample parking available in Ogmore-by-Sea. There is also a two-hourly bus service from Bridgend Bus Station that stops near the car park.

START

The Rivermouth car park in Ogmore.

CONTACT DETAILS

Southern Wales Tourist Information, Caerphilly Visitor Centre, Twyn Square, Caerphilly, Gwent CF83 11XX
t: 02920 880011
f: 02920 860811
e: info@southernwales.com
w: southernwales.com

Ordnance Survey Explorer Map number 151
© Crown Copyright 2008

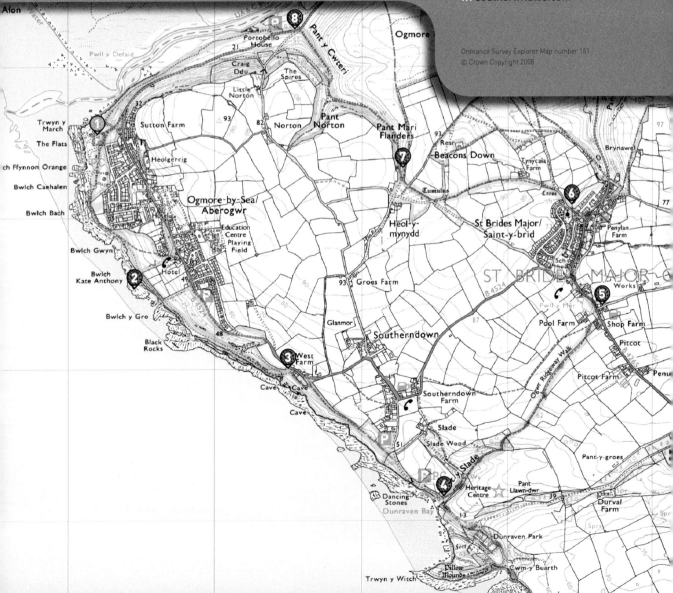

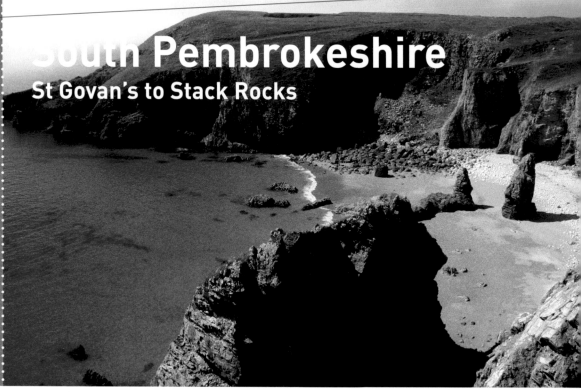

South Pembrokeshire
St Govan's to Stack Rocks

MEDIUM

ACCESS

7 MILES

3:00

ST GOVAN'S RIBS

On the east wall of St Govan's Chapel a gap leads to a small cleft in which legend has it that St Gobham used to hide. Once inside, the magical cleft closed, protecting him from his enemies. When they left it opened again, but you can still make out the imprint of his ribs.

This coastal walk across beautiful grasslands with excellent scenery and wildlife takes you through some of the highlights of the Pembrokeshire Coast National Park.

This part of the coast is very different from the north Pembrokeshire walk. Instead of walking through a hilly landscape of hard volcanic basalt you walk through a landscape made from limestone. The rock was formed from the sedimentation of microscopic skeletons and small shells of marine animals about 350 million year ago, and it is relatively soft. It is also heavily fractured, both vertically (called **joints**) and horizontally (called **beds**). Over time the sea has forced its way into the beds and joints, and eroded the rock to form some amazing features. The soil on top of this rock is thin and poor, but has created a rich and diverse wild flower population able to cope with the lack of nutrients. Combined with the low human development on the coast you may see birds and insects that you would not see anywhere else.

Starting in the car park at St Govan's, towards the sea at the front there is a gap leading to a footpath. Go through the gap, turn right and start to walk towards the gate with a small guard hut that marks the entrance to the Castlemartin range. Before you enter the range and start your walk to Stack Rocks, you can choose to visit St Govan's Chapel by following the path to the left, either now or on your way back.

② **St Govan's Chapel** was famous in medieval times for curing eye diseases and lameness. To reach it take the path signposted, turn left and descend a flight of uneven steps (the number of which is allegedly uncountable by mortals). The tiny chapel directly in front of you dates back to the sixth century. The small well in the chapel is where the healing waters were said to flow, but if you leave the chapel through the door in the west wall about 60 feet to the south is the remnant of the saint's holy well, which dried up last century. No one actually knows who St Govan was. The most likely suggestion is that he was St Gobham from County Wexford in Ireland, who was a contemporary of St David, – the patron saint of Wales. Another more exciting possibility is that Govan is a modification of Gawain – one of King Arthur's knights of the Round Table.

③ Returning to the main path and the gate to the firing range, you should check that no red flag is flying and that the signpost confirms that the path to Stack Rocks is open. Once through the gate follow this good path to Stack Rocks and the Green Bridge. Throughout the walk you will see small white stakes on the northern side of the path and it is very important that you stay to the seaward side of these. Within a few minutes the path comes close to **Stennis Ford** – the first significant coastal feature. This relatively narrow gap has impressive precipitous walls descending to

a small inaccessible beach and was probably formed by the sea eroding a bedding crack in the limestone to form a cave, and then the ford when the cave roof collapsed. Follow the path for another few minutes to reach Huntsman's Leap.

④ It's not hard to guess how **Huntsman's Leap** got its name. The apocryphal story is that a huntsman leapt the chasm on horseback and then died of shock when looking back at his achievement. Given that on very old maps the feature is referred as Penny's Leap and Adam's Leap it's not too likely. However, it remains an amazing 130-foot deep gash in the ground with walls that touch at its seaward end. It was almost certainly formed in the same way as Stennis Ford and both locations are great places to see rock climbers in action.

⑤ Follow the path enjoying the **wild flowers** and butterflies to reach a gate at a three-way junction. Go through the gate and over the cattle grid. The land that is left between the deep indentations is surrounded by water on three sides and provided the Iron Age inhabitants of the area with opportunities to build easily defended forts. If you walk a little way down the path heading towards the sea to the right of the fenced building you should be able to make out a man-made ditch and bank system that protected the inhabitants of what is called the Castle. With a stunning natural arch on one side it must have been a good spot and excavations here in the earlier

CLIMBING

With so many vertical rock faces this region is one of the best sea cliff climbing areas in the UK. Usually you cannot get good views of people climbing without binoculars, but the proximity of Huntsman's Leap and Stennis Ford to the coast path means that you can get some amazing views of rock climbers at play.

CLASSIC LIMESTONE SINK-HOLE

Back at the path a few hundred yards past the Castle there is distinct circular dip in the ground to the right that looks quite unusual, given the relatively flat grassland around. This is actually a classic sight in limestone country. Over time, beneath the soil, water has seeped along the rock joints and beds, and eroded the rock beneath until eventually the surface has collapsed. Inside the sink-hole you can see that the shelter allows different plant communities to prosper.

BUMBLEBEES

Since the range was requisitioned in 1938 the region has been left as unimproved grassland used for winter sheep grazing. This management of the land means plants can flower and so it is perfect for bees. You can see up to ten species of bumblebee in the range in what is one of the largest and best habitats in the UK.

half of the 20th century produced both Roman and British ceramics.

6 Continue along the path and after crossing a second cattle grid and passing through a gate, about 200 yards onwards towards the sea is a fine example of a **blowhole**. If you want to look, take care of the drop and leave the path to head towards the sea. The blowhole is the landward end of a sea cave and in very bad weather the sea is forced up this cavity. Eventually the cave will collapse to leave another Stennis Ford, but for now the blowhole provides shelter for ash and blackthorn. The path continues past the blocked door and windows that make up the ruins of a 19th-century dwelling before circumnavigating Bullslaughter Bay. The path splits here, but both branches join up just before another gate and cattle grid on the other side of the bay. Among the wildflowers here you should keep your eyes open for bumblebees. As you walk onwards look inland. Behind the buildings that make up Merion military camp you can see the prominent steeple of Warren church. This once derelict building was restored in 1989 and is used by the local community and the army. The spire is so obvious that it can be used as a navigation mark for shipping out in the Bristol Channel.

7 You will soon reach another Iron Age fort called Crockeydam Camp, where excavations again revealed Roman and British pottery as well as a partial skeleton. Continue past the sandy Flimston Bay to the prominent, heavily eroded headland that is the Iron Age fort called Flimston Bay Camp.

As you continue on your way you will start to notice vast numbers of seabirds in the sky. You must be careful of the drop here because you have reached the amazing pinnacles of

the **Elegug Sea Stacks**. Elegug is Welsh for guillemot and huge numbers of them nest here and colour the rock white with guano. You will also see razorbills, fulmars, kittiwakes gulls, and possibly the black, crow-like plumage and red beak and legs of the relatively rare chough.

8 Once past the stacks you leave the firing range through a gate and reach the car park and picnic area of Stack Rocks. It is a great place for lunch. Be careful of the sharp drops as you walk towards a ramped viewing platform. From here there are amazing views of the limestone arch of the **Green Bridge of Wales**. This feature was formed by the destructive power of the sea enlarging the joints until eventually rock fell away to leave the arch.

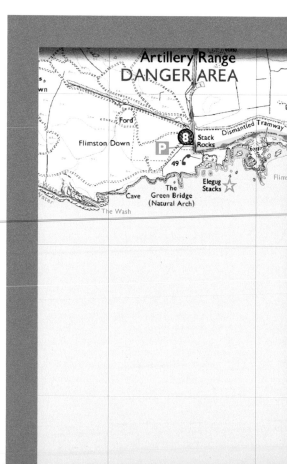

In time, of course, the Green Bridge will collapse and leave another stack pointing to the sky. Once you have finished you can go back to your start point at St Govan's by the local bus from the car park, or return along the path you have just followed.

South Pembrokeshire

ADVICE

The majority of the walk is within Range East of the Army Training Estate Castlemartin Range. This means access can be restricted due to live fire exercises, generally between 9.00 a.m. and 4.30 p.m. on weekdays, and occasionally 7.00 p.m. to midnight for night exercises. You should phone the number in the contact details below for the latest information and must not enter the well-fenced range area when red flags are flying. When you enter the firing range you must keep to the seaward side of white posts on the northern boundary of the path and not pick up any metal objects. In places there are very sharp cliff-edges. The nearest facilities along this walk are in Bosherston.

PARKING

This walk can be done in either direction and there are pay-and-display car parks at both ends of the walk. Both ends are also served by buses that run from nearby Pembroke.

START

The walk starts in the car park at St Govan's. You can still visit St Govan's Chapel when firing is taking place.

CONTACT DETAILS

Pembrokeshire Coast National Park Authority, Llanion Park, Pembroke Dock, Pembrokeshire SA72 6DY
t: 0845 345 7275/01646 689076 (24-hour answerphone: details of firing exercises 01646 662367)
e: info@pembrokeshirecoast.org.uk
w: pcnpa.org.uk

Ordnance Survey Explorer Map number OL36
© Crown Copyright 2008

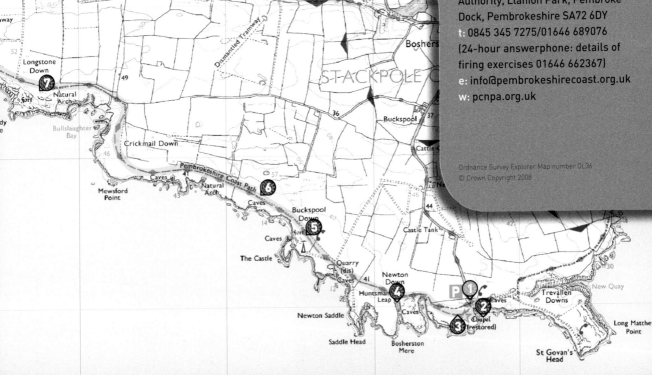

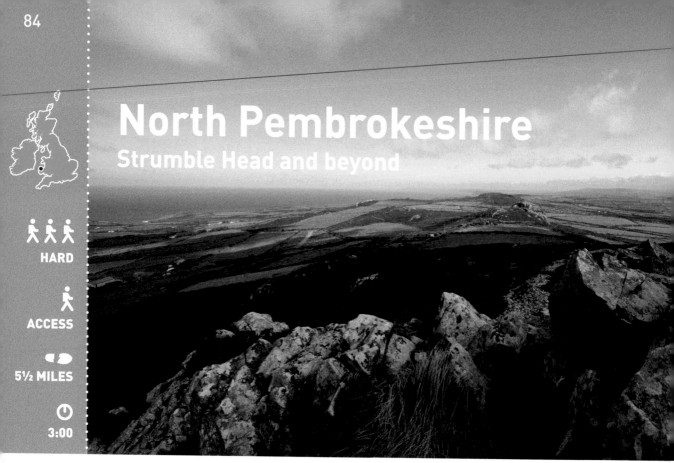

North Pembrokeshire
Strumble Head and beyond

HARD

ACCESS

5½ MILES

3:00

This circular walk, often across challenging terrain, includes an ascent of over 2000 feet, but the walker is rewarded with some stunning sea views and a chance to see some of Britain's rarer wildlife.

. . . .

GREY SEALS

Grey seals are quite common on this coast as there are two large colonies on Skomer and Ramsey Island to the south-west. They are big animals – males can be over 6 feet in length with females slightly smaller. They breed on the local beaches in autumn and you can often hear the pups crying out.

❶

Route-finding during this walk is relatively simple. If you parked on the side of the road, start by walking towards the **lighthouse**. Strumble Head is a popular spot and justifiably famous as a viewpoint for wildlife. The rough rocky coastline is made

from relatively hard rock called basalt, which was formed when molten lava first reached the atmosphere about 450 million years ago. The lava crystallised rapidly and formed the grey-black rock you can see all around you. This is igneous rock named after the Latin *ignis*, which means fire. The hard igneous rocks led to the formation of this rocky coastline, which not surprisingly has a well-deserved treacherous reputation. There were more than 60 ships lost in the region in the 19th century and when Goodwick harbour (a few miles to the east) was developed at the beginning of last century Strumble Head Lighthouse was built in 1908 on Ynys Meicl (St Michael's Island). Its importance can be seen by the fact that it cost £40,000 to build (about £2½ million in today's money). It went fully automatic in 1980 and so there is no lighthouse keeper. You can go inside a few times a year, though, on well advertised open days. At the bottom of the car park there is a bridge crossing over to Ynys Meicl and in the waters below you can often see grey seals.

2 At the back of the car park, follow the signpost westwards and after about 400 yards you come to the first of many stiles. The path weaves around the edge of **Carreg Onnen Bay**, and after another couple of stiles and a short flight of steps you'll reach a bench with magnificent views back towards the lighthouse. Continuing along the path take a moment to enjoy the pinks, reds and purples of the heather and bright yellow of the gorse. These are perfect robust plants for the coast and they provide a wonderful sheltered habitat for birds, insects, many species of butterflies and moths, and, of course, snakes.

3 As you follow the path and cross a small stream you can see the little cove of **Pwll Arian** (Silver Cove) down to your right. There are nesting kittiwakes on the rocks on the edge of the cove. As the path continues it enters a pair of boggy areas separated by a stile. The route is less defined here as people have tried to stay out of the mud, but if you stay out to the right you will get better sea views and the paths join up further along

where the ground dries out. Eventually, with a fence on your left and a sign pointing onwards, the path virtually takes a 90-degree left turn at March Mawr. In good weather you should be able to get your last view of the lighthouse as the path turns to the south-east. Instead of turning left, if you continue straight on for about 50 yards there is a very good south-facing terrace amongst the heather – it's an excellent lunch stop with views all the way from the lighthouse in one direction to the whole of St David's Peninsular in the other.

4 The main path stays close to the fence and leads uphill past some interesting quartz mineral bands in the rocks underfoot. After a while you pass some rather out of place brick buildings on either side of the path left by the Ministry of Defence (MOD) after the last two wars. Continue onwards to the small square bay of Porth Maenmelyn. As you follow the path around the bay, go through a gate and then for the first time temporarily leave the volcanic basalts behind and step onto a much softer and easily eroded shale rock. The basalt continues on the other side of the bay. Follow the path around to magnificent views of a Stone Age hill fort on **Dinas Mawr**, with Pwll

BE ALERT – FOR SNAKES

If you are lucky and extremely quiet you may see an adder – Britain's only poisonous snake. They are easily identified by their bold zigzag pattern and are not aggressive. The heather and gorse provide a perfect habitat with good feeding and shelter. They try hard to avoid human beings and are protected by law against being killed or injured.

INVADERS MEET THEIR MATCH

The last invasion of Britain was in 1797 during the Napoleonic Wars. An Irish-American Colonel called Tate landed with 1400 men from French warships with a view to marching through Wales to attack Liverpool. The invaders looted the local area and set fire to the church in Llanwnda before being routed and captured by local yeomanry and villagers after only three days. There is a beautiful tapestry commemorating the invasion on display in Fishguard Library.

Deri as a backdrop. It must have been easy to defend Dinas Mawr and now the steep walls provide good bird nesting sites. You are certain to see fulmars and perhaps even the relatively rare chough – a distinctive black bird in the crow family with a red beak. The path now winds upwards through a kissing gate and finally over a stile before reaching the Pwll Deri Youth Hostel. There is an outside tap here so you can refill any water bottles you have before the final uphill segment of the walk.

5 From the youth hostel follow the drive out onto the road. It is worth taking a short diversion here to see a monument for the Welsh Poet Dewi Emrys (1879–1952). About 200 yards down the road on the right is a simple but striking monument. The small plaque on the base has a Welsh poem, the translation of which is, '*and these are the thoughts that will come to you when you sit above Pwll Deri*'. Looking past the monument at the 500-foot high cliffs you certainly do think. When you have finished, return to the youth hostel, perhaps pausing to rest on the benches at the viewpoint on the left.

With the youth hostel on your left take the middle of the three roads to Tal Y Gaer Farm. After about 60 yards you reach a junction and directly in front of you is a strange entrance to an underground **corbelled hut**. There is much dispute over the age and purpose of the hut, with suggestions ranging from a pigsty to a fourth-century hermit's cell, although it is similar to huts that have been found in Ireland. By the corbelled hut turn left and take an immediate right over the stile on a marked footpath up to the summit of Garn Fach – the hill rising above you. After about 50 yards the path splits, the branch on the right contours around the Garn, but it's worth taking the left branch straight up to the summit.

6 It is a short and brisk ascent to reach the Garn plateau. Once there look immediately to the left and see the effects of slower cooling on lava. As the molten rock cools it contracts and develops fractures,

which under certain conditions can join and form a regular pattern. This is **columnar basalt** and it is exactly the same as that found at the famous Giant's Causeway in Ireland. Continue climbing until you reach the trig point on the very summit and enjoy the wonderful views. Garn Fach was an extensive Stone Age hill fort and from here you can see the remains of stone ramparts several feet high. Not faring so well just to the west are the remains of a First World War coastal observation post. The view to the west towards Pwll Deri and St David's is stunning, while to the north is the lighthouse where the walk started. To the east you can see almost the entire battlefield from the last time Britain was invaded.

7 From here you will be delighted to know that the walk is almost entirely downhill. From the summit head east on a well trodden path down to the road. Turn left and follow the road down through rich farmland and a **small farm** before you reach a T-junction. Take a left and follow the road back to Strumble Head, passing through another small farm, and after a few twists and turns you will reach the cattle grid mentioned at the very start of the walk.

8 As you walk along the coastal road back to the lighthouse there are two more treats to enjoy. The ex-MOD brick building down to the right is a great wildlife spotting point. There is good shelter and wildlife identification boards inside, and you never know what you may see. Just to the east of this the rocks near the sea have a rounded,

bulbous shape. This is one of the best outcrops in Wales of what is called pillow lava, which is formed when molten lava erupts beneath the sea and cools extremely quickly. When you look around at this landscape it's strange to think that it has been formed mostly by volcanoes.

North Pembrokeshire

ADVICE

This walk is both difficult and potentially dangerous. It is difficult because it covers some awkward terrain and has a surprising amount of ascent. It is potentially dangerous as it involves walking on generally unfenced coastal paths with occasional large drops. At all times you are strongly advised to keep to the well-trodden pathway. There are many spots for a rest and it is possible to find shelter from the wind whatever its direction. There are no facilities along this walk.

PARKING

Just after crossing a cattle grid the road to Strumble Head reaches the coast and takes a sharp left turn. Cars can be parked on the seaward side of the road or in a free car park a couple of hundred yards further on. There is also a bus service to and from Fishguard.

START

The walk starts on the cliff-top opposite the lighthouse.

CONTACT DETAILS

Pembrokeshire Coast National Park Authority, Llanion Park, Pembroke Dock, Pembrokeshire SA72 6DY
t: 0845 345 7275/01646 689076
e: info@pembrokeshirecoast.org.uk
w: pcnpa.org.uk

Ordnance Survey Explorer Map number OL35
© Crown Copyright 2008

Llyn Peninsula
A dramatic and dangerous coastline

MEDIUM

ACCESS

6 MILES

4:00

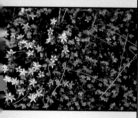

WILDFLOWERS

The relatively impoverished soils of the headland offer ideal growing conditions for wildflowers. There is lots of low-growing wild thyme, which has beautiful purple flowers in summer, and English stonecrop with tiny pale pink flowers and succulent leaves.

The Llyn Peninsula can be visited all year round and offers carpets of cliff-top flowers in spring, fabulous sunsets in summer, and wild and stormy seascapes in autumn and winter.

This part of the Llyn Peninsula showcases some of North Wales' most beautiful and unspoilt beaches, including St Tudwal's Road beach at Abersoch, Porth Ceiriad and the spectacular Porth Neigwl or Hell's Mouth, where the walk commences.

1 Start the walk by leaving the car park at **Hell's Mouth** and following the clear path through the dunes. When you get to the beach, pause for a moment to soak up the majesty of this huge 4-mile long curving bay. The location is popular with surfers, so if there are waves there's bound to be lots of activity on the water.

The Welsh name for the beach, Porth Neigwl, may derive from the name of 11th century local landowner Nigel de Loryng, who aquitted himself well in the Battle of

Poitiers in Gascony and was awarded a lot of land in this area by the Black Prince, eldest son of King Edward III of England and father to King Richard II of England.

When Nigel de Loryng was granted lands here there was a township, or collection of scattered farmsteads and small holdings, in the Hell's Mouth area. The only evidence of that township remaining today is the farm, Plas Neigwl, just inland from the beach, which would have been the centre of the township.

The English name, Hell's Mouth, is a much clearer reference to the fact that this is one of the most treacherous stretches of coast in the whole of Wales. Literally hundreds of vessels and cargoes have been lost in the jaws of Hell's Mouth. One such wreck, during the First World War, saw 17 victims buried in a mass grave in nearby Llanengan village church.

Apart from just bad weather, there was also a more sinister side to some of the wrecks at Hell's Mouth. In 1629, bonfires set on the mountain of Rhiw were used to lure a French ship onto the rocks at the north-west end of Hell's Mouth. On board the vessel were several French nobles, decked in fine clothing and jewellery. The story has it that the wreckers boarded the vessel as it ran aground and went on a killing spree, cutting off fingers and ears to obtain jewels. Two of the wreckers were subsequently hanged by the neck until they were dead.

2 Continue by walking left along the pebble and sandy beach. As you go, notice the huge variety of brightly coloured pebbles on the beach. These have been washed out of the boulder clay cliffs at the back of the beach. Boulder clay is a glacial deposit that was laid down at the end of the last ice age as the glaciers melted and retreated. Much of lowland England and Wales is covered in boulder clay, which, as the name suggests, is a chaotic mixture of boulders and clay.

The boulders have been picked up and transported far and wide by glaciers, which explains the wide variety of rock types, as well as the rounded nature of the boulders.

The clay is known as 'rock flour' and is the result of the erosive power of the glaciers grinding away at the bedrock as they advanced and retreated. The boulder clay cliffs are unstable and are being eroded and undercut at the remarkable rate of a foot per year.

Walk up the low bank at the south-eastern end of the beach, marked by a large wooden marker post, and continue along the grassy embankment at the back of the beach.

3 Climb the stile at the very end of the beach and follow the path up to the left, skirting a low grassy hill. The path is waymarked as the **Llyn Coastal Path**. Cross over another stile into a field and bear to the right up a short steep hill. Pass to the right of the house called Greenland, following white waymarked arrows on the gate posts. Continue following the track across a cattle grid then, between Pen-y-Groes and a pond, cross the farm track and head for the next waymark.

At the next, larger, pond turn right. You are now on National Trust-owned designated access land. Walk towards the headland for dramatic views of Hell's Mouth. Continue to follow the coast path waymarks to the left. Climb the gentle incline past a shallow quarry to a trig point.

In spring and summer look out for carpets of wildflowers; yellow gorse, heather, pink thrift, stonecrop, blue flowered sheep's bit, yellow and orange bird's foot trefoil and wild thyme.

SHIPWRECK AHOY!

The list of wrecks on or around Hell's Mouth is a long and doleful one, and a large number of lives and valuable cargoes have been lost here, including cotton, tobacco and even gold. In 1878 a schooner called the *Twelve Apostles* ran aground in bad weather and a telegram that was sent out to the insurers during the event has since entered local folklore. It read: 'Twelve Apostles making water in Hell's Mouth'. It was not until the crew had been safely recovered that the quite unintentional humour of this missive was appreciated.

MERMAIDS' PURSES

On the beach look out for mermaids' purses. These are the discarded egg cases of skate. There are 15 species of skate and rays living around the UK coast and many of them are under threat. Rays give birth to live young, while skate give birth to young in egg cases. If you discover an egg case on the beach, the Shark Trust can help you identify it: eggcase.org

OPEN-ACCESS LAND

The open land in England and Wales to which people have a right of access is now shaded yellow on Ordnance Survey maps. Obviously, when on this type of land you should follow the Country Code. You are allowed to walk or climb, but not to cycle or camp overnight. Dogs are permitted, but must be kept on a lead of no more than 6 feet between 1 March and 1 July.

4 Follow the coastal path waymarkers around **Cilan Head**, keep seaward and don't follow farm tracks inland. Steep grassy slopes drop down on your right-hand side. These lead to very steep rotting shale cliffs, the haunt of only the most adventurous rock climbers, so are best avoided by walkers.

Cilan Head is an area of common ground covering 215 acres and there are grazing rights for 900 sheep, 79 cattle, 30 geese and three ponies on the headland. Evidence of very early settlement by man has been found here in the form of Mesolithic microlith tools – saw-like blades consisting of very small, sharp, stone flakes set in pieces of wood. These Mesolithic people were probably nomadic hunters and have left no evidence of their dwellings. It seems likely that they visited this site only sporadically, probably in summer. The various ponds dotted over the headland may have provided them with fresh water.

5 As you turn round the headland you are rewarded with glimpses of St Tudwall's Island West and its impressive lighthouse, built in 1877. Now you can enjoy really fine views of Porth Ceiriad, another stunning

sandy bay with steep cliffs behind. Atop these is Pared Mawr, an Iron Age hill fort, though little can be seen from this distance.

You soon enter a narrow, fenced-in permissive **path above Trwyn Llech-y-doll**, before descending to a small cove. Cross a bridge over a stream and climb steeply up the headland beyond. The Lleyn coast path detours inland here and you continue to follow its waymarks away from the coast, through several fields, to Muriau Farm.

6 Take the farm track on the left until you meet the road. At the road turn right and follow it for about 400 yards (ignoring the left turn) until you reach an old chapel.

7 At the public telephone and post box on the corner, turn left and follow the footpath sign down the side of the chapel, through a small gate, into fields. Cross a couple of fields, keeping to the right-hand side, until you come to a **farm track**. Cross this, go past the front of a house and head towards Hell's Mouth beach, which is in front of you. It does look like someone's garden, but it is the right path and there's a wooden gate in the corner of the garden.

8 Walk about halfway down the next field until you find a rusty iron gate in the wall.

Go through this and turn left down a farm track. Keep right of the farm house and follow the prominent fence line path with Hell's Mouth in front of you and a shallow valley to the left. Go through a rusty five-bar gate into a cattle field and keep left above a steep-sided valley with a stream. Follow this until you are able to descend to a wooden bridge and re-join the coastal path. Take this back to the beach and subsequently Hell's Mouth car park.

Llyn Peninsula

ADVICE

This walk isn't suitable for wheelchairs or buggies. Access to the beach is limited at high tide, so please check tide tables. Most of the walk is on land that is designated as 'open-access land' and owned by the National Trust. There are some sections, clearly signed, which are on 'permissive paths' by agreement between Gwynedd County Council and the landowners. Close supervision of young children is recommended on the cliff-top sections of the walk, be wary of weaver fish if paddling barefoot and dog walkers should carry a lead as there are sheep on the headland.

PARKING

Park at the free car park at Porth Neigwl, otherwise known as Hell's Mouth, at the south-eastern end of the beach, just outside the village of Llanengan. There is usually a refreshment van parked here, but there are no toilet facilities at this car park or anywhere else on the walk.

START

The path from Hell's Mouth car park is obvious.

CONTACT DETAILS

Abersoch Tourist Information Centre, Abersoch, Gwynedd LL53 7EA
t: **01758 712929**
f: **01758 712929**
e: **enquiries@abersochtouristinfo.co.uk**
w: **abersochtouristinfo.co.uk**

Ordnance Survey Explorer Map number 253
© Crown Copyright 2008

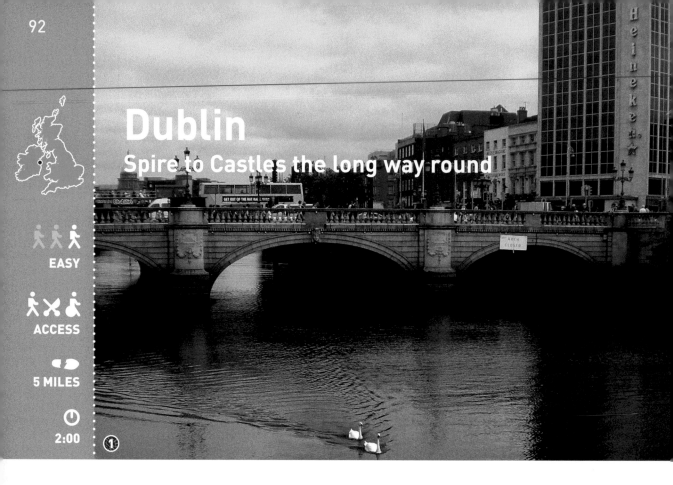

Dublin
Spire to Castles the long way round

EASY

ACCESS

5 MILES

2:00 ❶

THE GRAND CANAL

The main line of the Grand Canal is 82 miles long and links the Liffey with the Shannon. It was built between 1756 and 1796 and carried commercial traffic until 1960. As well as transport, the canal was also important in the development of Dublin's water supply and still supplies water for the lakes and fountains in St Stephen's Green, through which you'll be walking later. It also supplies water for the Guinness Brewery.

This walk takes you through the history, past and present, of one of Europe's most convivial cities.

❶ Start by the Spire of Dublin (officially the Monument of Light: *An Túr Solais*). This was completed in 2003 as a replacement for the Nelson Pillar, which was blown up in 1966. It is 393 feet high and is believed to be the tallest sculpture in the world. Walk down O'Connell Street towards the **River Liffey**. Although the street was originally laid out, as Sackville Street, in the 18th century, most of the buildings date from the 1910s and 1920s. Many buildings were seriously damaged or destroyed in the 1916 Easter Rising, its aftermath or the civil war.

On the right is the General Post Office, the site of the Easter Rising. All but the facade of the original 1818 building was destroyed:

the new section behind the facade was started in 1924 and opened in 1929.

Keep going past the statues of Jim Larkin (1876–1947, trades union activist), Sir John Grey (1816–1875, chairman of the Dublin Waterworks Committee), William Smith O'Brien (1803–1864, leader of the Young Ireland movement) and Daniel O'Connell (1775–1847, political leader and campaigner for Catholic emancipation), after whom the street was renamed in 1924.

❷ At the bottom of the street, turn left onto the quayside (Eden Quay) and head east along the north bank of the Liffey, crossing the road at the Butt Bridge, built in 1832, and going under the Loopline Railway bridge, built

in 1891. Then cross the road over the Talbot Memorial Bridge, built in 1978 and named after a famous temperance campaigner.

The riverside is now Custom House Quay and the Custom House itself is the impressive neoclassical building on your right. It was burned down in 1921 and subsequently restored, which is why two colours of stone can be seen. Shortly afterwards are the **Famine Sculptures**, a memorial to the Great Hunger (*An Gorta Mór*) of 1845–1849, in which around a million people died. Although the quayside is now separated from the river by a substantial wall, paired mooring hooks for ships can still be seen at regular intervals.

3 Pass the rolling bascule bridge (now fixed) across the entrance to George's Dock and then cross the Liffey by the modern Sean O'Casey Memorial Bridge. Turn left along **City Quay**. This area is undergoing considerable redevelopment, but there are still signs of its former life as a busy port. On your right are various shipping company offices – those of the Tedcastle Line and British and Irish Steam Packet Company Ltd are still clearly labelled. On the right are warehouses and the remains of the narrow gauge railway system that served the quays.

Further along (Sir John Rogerson's Quay) is a diving bell, designed by the splendidly named Bindon Blood Stoney. He was the engineer responsible for much of the docks development in Dublin and the diving bell was used to build the North Wall Extension.

4 Take the first right after the diving bell, Forbes Street, which leads to the **Grand Canal Docks**. They were built between 1791 and 1796, to replace an earlier basin to the west, and cover 25 acres of water. You now follow the canal for a couple of miles as it curves around south Dublin.

Misery Hill – on your right as you reach the Grand Canal Docks – was once an execution site, where miscreants were executed and left to rot in chains as a deterrent to others. Nowadays it offers nothing more exciting than some bland corporate sculpture.

5 From the quayside at the larger of the docks cross Pearse Street to Grand Canal Quay. The street name is misleading, as there are buildings between it and the quayside for almost its whole length, but there are occasional glimpses of a pleasantly mixed set of warehouses on the opposite bank. Access to the Waterways Visitor Centre is on your left. Pass under an extremely low railway bridge, which has platforms for the DART Grand Canal Dock on top of it, and when you reach Grand Canal Street turn left and then right to join the towpath beside Clanwilliam Place.

Stroll along the towpath, past McHenry Bridge, to **Huband Bridge**. A short distance down Mount Street (on your right) is

ST STEPHEN'S GREEN

Until the 17th century this was an area of marshy ground outside the city of Dublin. In 1664 it was turned into a walled park and the land around it sold off for building by the Dublin Corporation. In due course it became a prosperous enclave, with the park kept for residents' use. It became a public park in 1877 at the instigation of Sir Arthur Guinness, of the brewing family, and was laid out in more or less its current form at his expense.

St Stephen's Church, known as the Pepper Pot. The original rectangular building was completed in 1824 and the curved apse was added in 1852. Sir Charles Villiers Stanford, the composer, lived nearby as a child and is said to have taken his first music lessons in the church.

Shortly after Huband Bridge there is a statue of the poet Patrick Kavanagh on a bench – a reference to one of his best known poems.

The next bridge over the canal is officially MacCartney Bridge, but it's normally known as Baggot Street Bridge. There is a cluster of shops, cafes and – inevitably – pubs on the south side of the bridge, so this is an excellent place to take a break, with the walk half completed.

6 **Eustace Bridge**, the next bridge, also carries water pipes over the canal, with rather elegant cast-iron bleed valves – similar in function to those on central heating radiators – on top. After this the towpath is briefly, and for the first time, separated from the road by buildings (a row of cottages).

Go under the Luas (tram) bridge and station, and shortly afterwards turn right down Charlemont Street. At the end turn right into Harcourt Road and almost immediately left into Harcourt Street, with the Luas line now at ground level. The impressive building on your right, by the Luas stop, is the former Harcourt Street Station, the terminus of a suburban line to Bray. The line closed in 1958, but was reused for the Luas Green Line, opened in 2004, running at street level to St Stephen's Green instead of the original terminus.

Pass the Harcourt Hotel, which includes a home of George Bernard Shaw, and follow the Luas tracks to St Stephen's Green. Enter the park through the first gate you come to (the south corner) and, having explored as much as you feel like, leave by the far gate on the same side (the west corner).

7 Head down Grafton Street, which is pedestrianised, until you come to the famous Bewley's Oriental Cafe on your left. Joshua Bewley, a Quaker tea merchant, started the company in 1840 and the Grafton Street cafe opened in 1927. Immediately after Bewley's turn left into Johnson's Court, pass St Theresa's Church on your right and at the far end of Johnson's Court turn left and immediately right across Clarendon Street into Coppinger Row.

When you emerge into South William Street the frontage of **Powerscourt House** (now a shopping centre) is on your right and directly opposite it is the pedestrianised Castle Market. Walk along here, cross Drury Street and enter the South City Markets. These were built in 1881, but what you can see mostly dates from reconstruction after a fire in 1892.

8 Leave the market at the far end and turn right into South Great George Street and left along Palace Street – it's rather scruffy. At the end of Palace Street admire the splendid building of the **Sick and Indigent Roomkeepers' Society**. It's the oldest charity in Dublin and was, as the building says, founded in 1790. It is still active, but despite the signs now operates from an office on Lower Leeson Street, as the original building would have been too expensive to convert to meet modern standards. It's now a private house.

Turn left via Palace Hill Gate into Dublin Castle and explore as much as you like: the walk ends here. If you would like to return to the starting point, leave by the Cork Hill State Entrance and bear right past City Hall before turning left down Parliament Street. Cross the Liffey by Grattan Bridge and turn right along the walkway, past Ha'penny Bridge, to the foot of O'Connell Street.

Dublin

ADVICE

Dublin is an easy and pleasant city to visit, and one thing's for sure, you won't have trouble finding places to eat and drink on your walk.

PARKING

As with any city there are parking restrictions during business hours, but the city has a number of pay-and-display car parks. Consult a local map for details.

START

The starting point is near Connolly Station and the Red Luas line, and the end point is only ten minutes away from the start.

CONTACT DETAILS

Dublin Tourist Information, Tourism Centre, Suffolk Street, Dublin 2
t: (within Ireland) 1850 230 330
(within the UK) 0800 039 7000
w: visitdublin.com

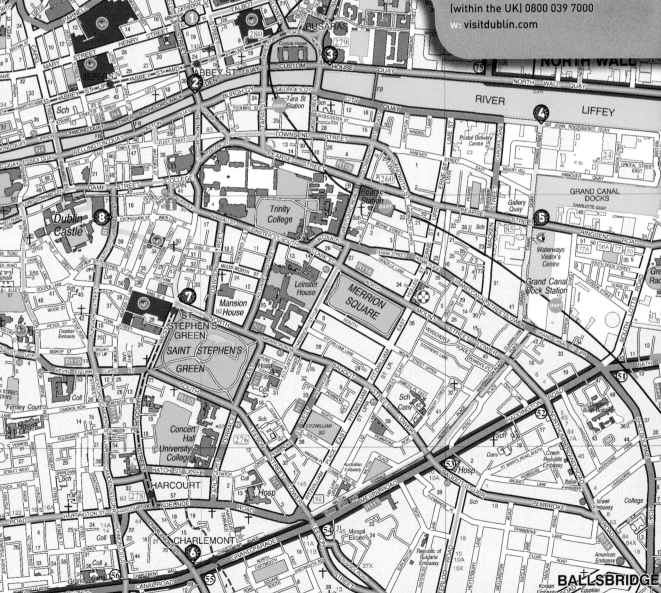

Achill Island
Mountains, mists and cliffs in the far west

HARD

ACCESS

5 MILES

4:30

This short but energetic walk takes in the spectacular Atlantic cliffs of Achill Head, as well as various human constructions in and around Keem Bay in Ireland's County Mayo.

1 Start in the car park at **Keem Bay**, which has a strong connection with fishing. It was the location for much of Achill's shark fishing industry during the 1950s and 1960s. At that time the basking shark was a frequent visitor to the waters around Keem Bay and it was hunted for its liver oil, which was exported from Achill to provide fine grade lubricant for the aerospace industry. A couple of canvas-covered curragh boats, used for this type of fishing, are still based here. Basking sharks are now safe, though, as they are a protected species.

Head west up the hill, keeping the turf wall about 3 or 4 yards to your left (it's boggy closer in). It is extremely steep in places: this first section climbs nearly 1000 feet at almost 45 degrees. About a third of the way to the top the slope lessens slightly. At this point cross the wall and head for the derelict coastguard lookout at the top of the hill.

As with much of the rest of the walk there is no single path, but it is easy to find a way up.

2 When you reach the top, take a few minutes – or longer – to recover your breath and enjoy the view. The large island to the south is **Clare Island**, at the entrance to Clew Bay. It was the headquarters of Grace O'Malley, the pirate queen, who also built the castle at Kildavnet near the southernmost point of Achill Island. Beyond Clare Island is Inishturk and beyond that Inishbofin. The prominent mountain to the south-east is Croagh Patrick, where St Patrick is said to have fasted for 40 days in the fifth century. At the end of his fast he banished all snakes from Ireland. Every year on Reek Sunday (the last Sunday in July) around 25,000 pilgrims climb to the summit at 2510 feet.

The walk now follows the cliff-tops to the north-west. The path is well worn and easy to follow, but it runs very close to a near-vertical drop of around 1000 feet to vicious rocks and the sea, so great care is essential. This is not really suitable for people with a tendency to vertigo. If the wind is strong and from the east then you are advised to turn back now.

3 The first cliff-top peak reached is at 1071 feet and the next (the highest) at 1089 feet. Blackrock Mayo lighthouse is now visible on a small island to the north. It was opened in 1864 and since 1999 has been solar powered: an array of 50 solar cells charge a battery bank which in turn powers the light. There are no resident keepers, but an attendant visits every three weeks to carry out maintenance.

Looking back to the cliffs along which you have walked, the contrast between the smooth slope to the inland valley, aligned with the rock strata, and the jagged vertical descent to the sea is very obvious.

4 Follow the cliff-tops as far as you can, until you are looking down on Achill Head. In the valley to your right you can see the remains of the abandoned booley village of Bunowna. To the east, the mountain immediately in front of you is Croaghaun, 2190 feet high. The cliffs to its north are claimed to be the highest sea cliffs in Europe.

Now head downhill to the inlet below the village, keeping about 10 yards away from the cliff edges. A steep gully opens up on your right: near the bottom follow the path into this gully and cross the stream at a single slab bridge by the turf dyke, and then head for the prominent quartz outcrop on your left. Several large caves are clearly visible further round the **coast of Achill Island**. They were popular with smugglers, which was the reason for the coastguard presence in Keem Bay.

5 When you get to the main stream, turn right and head up it, past the remains of the village. There were 17 houses in total, although only eight or nine still exist in any substantial form. After the last house, make sure you are on the left-hand (north) side of the stream and move away from the middle of the valley, which is very boggy. The going is easiest just below the rocky screes, although it can still be a little damp.

BOOLEYING

Booleying was the ancient custom of moving animals to summer pastures. The people attending them would live in booley villages. Since these were only for temporary use in summer the buildings were extremely simple: normally small, single-room, stone huts with turf roofs covered with thatch. Booleying was officially outlawed in 1697, but continued much longer in the west of Ireland, and longest of all on Achill.

CAPTAIN BOYCOTT

Captain Charles C. Boycott (1832–1897) moved to Achill in 1853, first to Keem and later to Corrymore House. In 1877 he went to Ballinrobe on the mainland, where he worked as land agent for Lord Erne. In a dispute with the Land League over the treatment of tenant farmers, he was ostracised: no locals would work for him, talk to him, serve him in shops or sit near him in church. As a result of wide publicity at the time, his name entered the English language for being so treated.

PENAL TIMES

From the 1690s onwards a series of draconian laws were passed with the aim of suppressing Catholicism in Ireland and encouraging Catholics to convert. Catholics lost voting rights, were taxed more heavily and were forbidden from educating their children (in Ireland or abroad) and from entering most professions. Reform came slowly, starting in the time of George III, and was largely completed with the Roman Catholic Relief Act of 1829.

Pass the two small loughs and cross another turf wall. The ridges in the field here are the remains of lazybeds, the traditional cultivation system of the west of Ireland and Scotland. Raised beds were made of layers of peat and seaweed, and were used to grow mainly potatoes.

Head for the **ruined house** ahead of you – it was built by Captain Boycott in 1853, together with the large store building on the other side of the valley. Originally it had an 'L' shape, with an east-facing courtyard sheltered from the prevailing westerly winds. The remains of two large windows with spectacular views over Keem Bay can still be seen.

6 From Boycott's house take the original access track on your left, bearing right after a few hundred yards on the path past the **penal altar**. In Penal Times, the public celebration

of Mass was illegal – those attending were deemed to have rioted and could be publically flogged as punishment. The stone altar marks one of the sites on which travelling priests held services outdoors.

7 From the altar take the small path to the right and pass to the right of the large white building. This was built as a coastguard station around 1912, though it fell into disuse at independence. It was bought as a ruin in the 1940s and is now a private house. Past this building the path goes between large boulders, between a large outcrop and a smaller rock, and so down to the **shore**.

8 If you are feeling adventurous, and if the weather is good, you can return via the summit of **Croaghaun**. Details of the route can be obtained at the Walking Information Centre in Keel. It is extremely steep and should only be attempted in good conditions and with proper equipment. Orographic cloud can shroud the mountaintop in minutes, so be prepared to cut the walk short at any stage if this happens and have a safe route down planned in advance.

However you choose to finish the walk you will have to retrace your steps to get back to the starting point at Keem Bay.

Achill Island

ADVICE

You can get to Achill Island by public transport – Iarnrod Eireann trains run to Westport, from where there are bus services to Keel. However, you will need to get a taxi for the last couple of miles to Keem. Driving is the most convenient way to get there, crossing the swing bridge to the island. Keem Bay itself has no facilities, though you may get ice cream there on a summer's day. The rest of Achill Island has shops, garages, hotels and an abundance of guesthouses – it's a popular holiday area.

PARKING

There is a car park by the beach at Keem Bay.

START

You start and finish the walk in the car park at Keem Bay.

CONTACT DETAILS

Achill Tourist Information Centre,
Cashel, Achill Island,
County Mayo, Ireland
t: (within Ireland) 098 47353
(outside Ireland) 00353 98 47353
f: 00353 98 47353
e: info@achilltourism.com
w: achilltourism.com

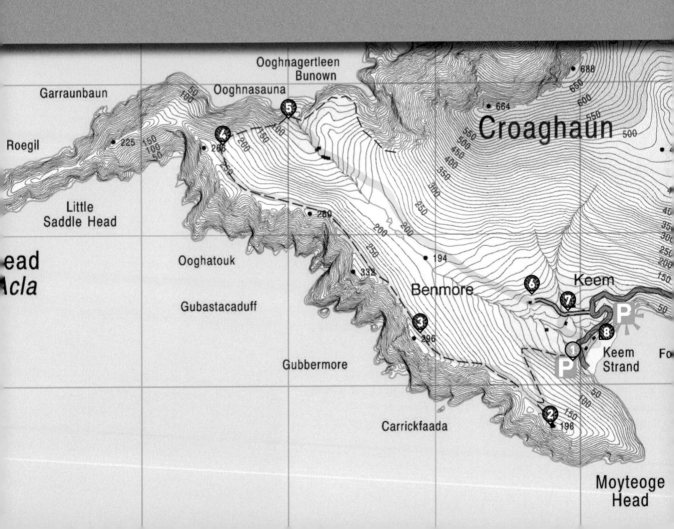

Anglesey
Land of the red squirrel

EASY

ACCESS

3 MILES

2:30

MARRAM GRASS

For hundreds of years Newborough Warren had an industry based on marram grass. Indeed, up until the last century, nearby Cwningar farm employed six women who harvested marram grass in summer and spent the rest of the year weaving it into mats and ropes.

The south-western corner of the island of Anglesey, or Ynys Mon, boasts the spectacular Newborough Warren dune system and magical Llanddwyn Island.

This walk takes you around part of Caernarfon Bay, with fabulous views of the Snowdonia Mountains, the Llyn Peninsular and the Menai Straights, and on to Llanddwyn Island itself, which has a fascinating social and geological history. The shore, dune complex and Llanddwyn Island are all part of Newborough Warren National Nature Reserve and are managed by the Countryside Council for Wales. Newborough Forest is managed by the Forestry Commision Wales.

① Start the walk by taking the **sandy path** from the car park which leads down to the beach. Turn right and walk along the beach by the edge of the sand dunes. Newborough Warren was named as a result of two quite different events. Newborough was

the result of King Edward I's decision to build a new castle at Beaumauris in the south-eastern corner of Anglesey and to relocate many of the villagers to a new borough. Warren is a reference to the thousands of rabbits that once inhabited the dunes. As they do today, although to a lesser extent, the rabbits fed on the succulent dune plants and the soft sand made for easy burrowing.

The dune system itself owes its origin to the legendary storms of 1313, when hundreds of tons of sand were swept inland, covering the fields and farmsteads around Newborough. Disaster struck because too many people were farming the fields, right down to a narrow barrier of dunes near the shore.

2 As you walk along the beach by the dunes, look out for marram grass and lyme grass, whose long roots reach down into the sand and help keep the dunes stable. Marram grass, in particular, has long been recognised as crucial to dune stability and as far back as Elizabethan times laws were passed to punish anyone who damaged this grass. During the Second World War, Newborough Warren was used for military training and as a result much vegetation was destroyed, and the dunes suffered badly from erosion. Local people helped to stabilise them by planting marram grass.

Other salt- and sand-loving specialists are common on the dunes, including sea sandwort, sea holly, sea rocket and biting stonecrop. Notice that many of these plants have succulent, rather than fleshy, leaves, which they use to store water, and thick skins which help to prevent water loss by evaporation and transpiration.

3 At the west end of the beach are three greenish coloured rocky outcrops and it's definitely worth stopping and having a closer look at them. These are **some of the oldest rocks in Britain** (as are the rocks on Llanddwyn Island itself) and they date back to at least the Precambrian period, 750 million years ago. They are also quite unusual in the way they were formed. Look closely and

you'll see that they have a lumpy, rounded appearance. This is what is known as pillow lava. It is formed when molten rock oozes out of undersea volcanoes. The blobs cool, break off and tumble down, forming pillow lava.

4 Continue across the narrow sand bar onto Llanddwyn Island. Here there is a stone shelter, which has a map of the island with all the pathways marked on it. There are many routes around the island, but it's certainly worth visiting Pilots Cove, the boathouse and the tower. Take the path which leads straight up the middle of the island, passing the remains of a chapel on your right and a **Celtic-style cross** on your left, dedicated to St Dwynwen, the Welsh patron saint of lovers.

5 Continue along the path until you reach a line of four tiny, **terraced cottages** with a cannon in front of them. Although today they are only inhabited during the

SOAY SHEEP

In 1956, the virulent disease myxomatosis virtually eliminated the rabbits in this area. As a result, more vigorous plants and grasses started taking over the dunes and Llanddwyn Island, which meant that many smaller herbs and short-lived wildflowers were crowded out. Consequently, in order to re-address the natural balance and preserve a range of species, an unusual and ancient breed of mammal called Soay sheep was introduced to graze on Llanddwyn Island. The Soay sheep are doing an excellent job of keeping the vegetation and shrubs down, and allowing the wildflowers and herbs to thrive.

Caernarfon Bay. Three pilots and two of their sons perished trying to help her.

6 Walk up past the boathouse to the whitewashed, circular tower at the end of the island. This tower was built in 1846 and had a light that sailors used to help guide them safely into the Menai Straights.

However, don't neglect to come down from the tower and explore the beach at its foot, as piles of **colourful shells** often accumulate here. If the tide is out you can also get a closer look at some of the ancient rocks of Llanddwyn. On this beach, in particular, if you walk around to the left, the sea-washed outcrops of rock resemble elaborate sculptures and they come in a quite startling range of colours, from green through to dark red and palest pink.

Leave the island by one of the shell-covered paths. Cross the narrow sand bar again and retrace your steps along the beach to the pillow lava.

7 In front of two rocky pillow lava outcrops, head left up the dunes to a **narrow sandy path**. Follow the path along the top of the dunes as it rises gently to a parking area. From here follow the wide forestry commission track into the trees, bearing to the right at the sharp bend.

Wheelchair and buggy users can enjoy a shorter walk through the forest, culminating with views out to Llanddwyn Island, by

HOME TO ROOST

Adult ravens live in pairs or as individuals, but juveniles congregate in the evening in roosts of up to 200 birds. It is thought they do this to exchange information about the location of carrion, which they would be unable to defend alone from pairs of adult ravens. The young ravens fly in just before dusk from as far afield as Ireland and the Isle of Man.

summer by volunteer wardens, these cottages were built to house the pilots who once lived on the island. The need for pilots was recognised because the shifting sandbanks and bars of Caernarfon Bay represented a major hazard to shipping. The pilots would row out to meet approaching ships and guide them into the then busy port of Caernarfon.

The pilots also crewed the lifeboat, which was housed in the boathouse on the beach below the cottages. The cannon used to be fired to summon help from the village. Over the years the men and women of Llanddwyn Island took part in many brave rescues, but they also experienced tragedies, such as the occasion in January 1876 when the schooner *Margaret* ran aground on a sandbar in

following this trail from the main car park and then retracing their steps, as described below.

Continue along the forest track, looking out for birds and wildflowers as you go. Newborough Forest was planted between 1947 and 1965 to protect the village of Newborough from wind-blown sand, and to provide timber and jobs. The trees are Corsican Pine, a species which is particularly effective in trapping and binding sand, and which can also thrive on sandy, salty soil.

8 The forest is rich in wildlife. There are birds such as goldcrest, chaffinch and greater spotted woodpecker to be seen, and Newborough also boasts one of Britain's largest raven roosts. In spring and summer many wildflowers can be spotted without leaving the track. Look out for pale pink common restharrow, yellow, sweet-smelling ladies bedstraw and blue-flowered viper's bugloss. There are also various types of orchid, including the pyramidal and common spotted orchid. As you make your way back to the main car park and the end of the walk, keep your eyes peeled for Newborough's most elusive new resident, the red squirrel. There has recently been an extensive cull of grey squirrels at Newborough Warren as part of a programme to re-introduce the native British red squirrel back into the area. If you're lucky you may just see one scurrying through the trees.

FLASH OF RED

Contrary to popular belief, native red squirrels were not driven out by non-native American grey squirrels. Rather, grey squirrels carry the parapox virus, which does no harm to them, but which decimated the red squirrel population. Once it had lost ground, the smaller, less robust red has been unable to regain it unaided.

Anglesey

ADVICE

Part of the route is on Forestry Commission tracks which are suitable for people using wheelchairs or buggies. However, there is no wheelchair or buggy access to the beach or Llanddwyn Island. The route is accessible all year round, but check tide tables as Llanddwyn Island is cut off from the mainland during high spring tides. Dogs are banned on the right-hand side of the beach and on Llanddwyn Island from 1 May to 30 September.

PARKING

This is an excellent route for families with young children. It is not too far and there are many points of interest along the route. From the village of Newborough, follow the signs for Llanddwyn and Beach. Enter Newborough Warren through a barrier, where there is a payment meter (parking charges apply). Follow the road down through the forest and park in the car park. There are toilet facilities and picnic tables, but no refreshments.

START

From the car park, head down towards the beach.

CONTACT DETAILS

Countryside Council for Wales, Maes y Ffynnon, Penrhosgarnedd, Bangor, Gwynedd LL57 2DW
t: 0845 1306 229
f: 01248 355782
e: enquiries@ccw.gov.uk
w: ccw.gov.uk

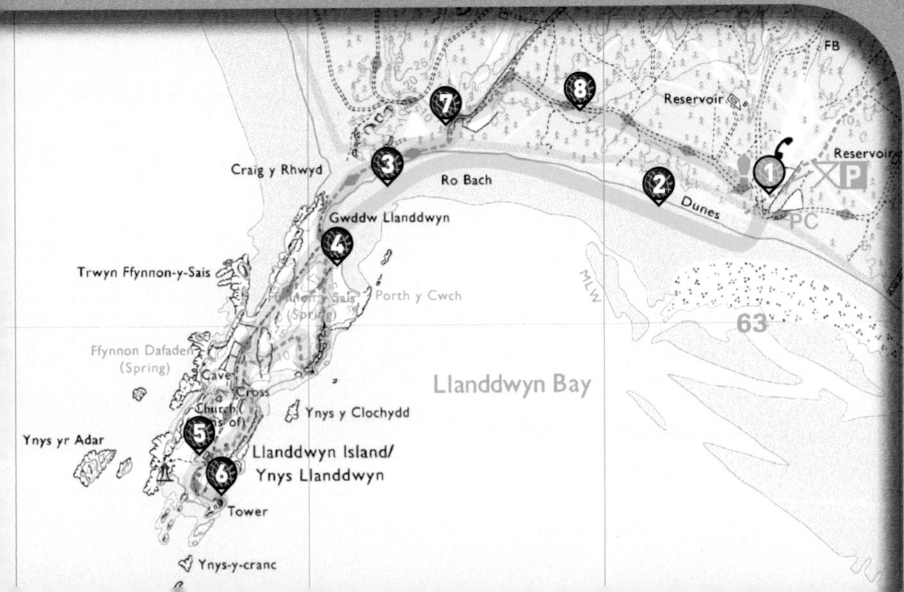

The Great Orme
A 350 million-year-old walk

MEDIUM

ACCESS

4½ MILES

4:00

LIMESTONE PAVEMENT

Limestone is a sedimentary marine rock. It is a very hard, but is susceptible to erosion by rainwater, which creates the classic features of limestone pavement. The free-standing slabs of limestone are known as clints and the deep cracks as grikes.

This walk offers stunning views of Llandudno Bay and the Great Orme, down the North Wales coast and across to Anglesey, and back up the Conwy River to Conwy Castle.

The dramatic limestone headland that forms the Great Orme is a distinctive part of the north Wales coast. There are excellent examples of limestone pavement on the plateau at the top and a variety of plants and insects associated with the lime-rich soil. The limestone of the Great Orme was laid down in the Carboniferous period about 350 million years ago, and is composed largely of the shelly and skeletal calcium carbonate remains of marine creatures. Look closely and you may find fossil corals and shells in the rock. The Great Orme is a country park and also a Site of Special Scientific Interest (SSSI).

The walk starts at the entrance to **Llandudno Pier**, an elegant and beautifully

restored iron structure dating from 1876. Steamers of the Liverpool and North Wales Steamship Company operated from this pier for many years. Continue along the road through a stone arch and enter the Marine Drive Toll Road. This spectacular corniche-style road runs all the way around the Great Orme. It was opened in 1878 to replace an earlier path, which Prime Minister William Gladstone complained about when he visited in 1868.

② Walk along Marine Drive enjoying great views of the steep and often overhanging limestone outcrops that form the upper and lower tiers of **Craig Pen-trwyn**. These are the Great Orme's premier rock climbing destinations, so don't throw anything over the seaward wall where climbers may be out of sight below!

Continue to stroll along the road, where there are a number of benches for resting and enjoying views of Llandudno Bay and across the bay to the Great Orme's smaller sibling, the Little Orme. On a clear day you will also be able to pick out a large offshore wind farm in the centre of the bay.

③ In spring and summer the limestone cliffs are decked with swathes of English stonecrop, pale blue harebells and even **orchids**. There are a number of arches in the cliffs where stone has been removed. Limestone from the Great Orme was used to build the suspension bridge at Conwy and was exported for use around Britain, too. Also keep a look out for the famous feral Kashmir goats of the Orme. They are said to have originated from a pair sent over from India as a present for Queen Victoria and have been roaming wild here since the 1890s.

At a fork in the road, walkers should follow the signs for the summit and visitor centre. At this point those with wheelchairs and buggies can continue along Marine Drive and complete a circuit of the Great Orme on this route.

Having left Marine Drive continue up the tarmac road around a couple of hairpin bends. Leave the road and follow a sign to the left up a grassy bank. Take the path to St Tudno's church. The entrance is down the road on the right. Note that there are public conveniences further down the road past the church, on the left-hand side.

④ The present **St Tudno's church** is a small stone building built in the 12th century. It's quite plain inside, but has a number of beautiful stained glass windows. However, none of these depict St Tudno and the only image of him is in the Parish Church of the Holy Trinity in Mostyn Street, Llandudno.

Virtually nothing is known about St Tudno's early life, but he is thought to have climbed the windswept slopes of the Great Orme in the sixth century to bring the message of Christianity to the people living in small stone huts on the headland. There are indications that in pre-Christian times the Orme may have been a centre of heathen worship, so that may be why he chose this location for his mission. There has been an active place of worship on the site ever since. St Tudno is the patron saint of Llandudno and the town's name means enclosure of Tudno.

ST TUDNO'S CHURCH

In January of 1839 a severe storm heavily damaged the church of St Tudno and for 15 years it lay open to the elements. However, after an appeal for £100, renovation began on St Tudno's day, 5 June, in 1855. The roof was repaired and the font, which had been taken away for use as a pump trough, was replaced. The church as it is today re-opened in October the same year.

FFYNNON POWELL

There is a legend attached to Ffynnon Powell spring, which is close to St Tudno's church. Folklore has it that this spring miraculously appeared after a local family of farmers named Powell prayed for God's intervention during a severe drought. They had been denied access to other wells in the area and needed a miracle to survive the drought.

HAPPY VALLEY GARDENS

The Happy Valley Gardens are in a dry and sheltered valley on the side of the Great Orme. The valley was formerly a copper mining site before becoming a stone quarry. In 1887 the area was landscaped and given to the town of Llandudno by local landowner Lord Mostyn, in celebration of Queen Victoria's Golden Jubilee. The gift is commemorated by a drinking fountain featuring a bust of Victoria.

Leave the church by one of the two up-hill gateways and follow the road around to the right. At the corner of the cemetery leave the road and follow a narrow track, again to the right, either through bracken in summer or across open ground for the rest of the year. Join a gravel track and walk around to the right onto the top of the Great Orme. The path becomes increasingly grassy as it crosses a plateau of meadows, limestone pavements and walls. In summer the heather is in bloom and the bright purple flowers form a stark contrast to the white limestone and blue sky.

Follow the path over the plateau keeping a high limestone wall on your left. At an obvious corner in the wall turn left and head out over an area of limestone pavement.

5 Here, beside a collapsed cairn, is a spectacular **view out to Anglesey** – Ynys Mon as it is known in Welsh – and Puffin Island or back inland to Conwy Castle and the Carneddau mountains. Continue to follow the limestone wall around the top of the Great Orme.

6 Ascend to the cable car and tram station, known as the **summit complex**. There are refreshments and toilet facilities available here. Alternatively, you can avoid the complex by keeping to the lower, well marked path that goes past Bishops Quarry.

The present summit complex, at a height of 680 feet, was built as a hotel in 1903. It is the terminus of a unique cable-hauled

tramway, which has carried visitors to the summit since that date, and of a cable car built in 1969. Both are still running today and offer an easy alternative route to the top. There was also a semaphore station here in the 19th century, one in a long chain built by Liverpool Dock trustees to transmit messages between Liverpool and Holyhead on Anglesey. Leaving the summit complex, follow the road downhill on a waymarked path by the side of the road, keeping the fence on your right and passing above the copper mine.

The copper mine on the Great Orme dates back 4000 years to the Bronze Age and is believed to be the most extensive copper mine of ancient times. The copper mining on the Orme reached its peak in 1830–1850 with over 300 men working in the mine. You can make a detour and visit the copper mine, which is open daily and offers tours around the workings. There is also a visitor centre with toilet facilities and refreshments.

7 Continue on down the track by the side of the road until just past the tram halfway station. Cross the tracks by a cattle

top of the artificial ski slope, before curving round to the right and descending gently via a series of well-spaced concrete steps into **Happy Valley Gardens**.

8 Go through the wooden gate into Happy Valley Gardens and descend through them via a series of winding paths. There are excellent views down onto the pier and **Llandudno**, and across the bay to the 464-foot-high Little Orme. At the bottom of the gardens go down a flight of steps back onto Marine Drive and walk right to return to the seafront.

grid in the road on a sharp bend. Head up right, towards a waymarked limestone path, keeping left at a fork in the path and passing under the cable car wires, along a green, grassy track. This takes you along past the

The Great Orme

ADVICE

This walk is accessible throughout the year. At least half of it can be done with wheelchairs or buggies and the rest can be completed by taking a slightly different route. There are refreshments and toilet facilities halfway round.

PARKING

Parking is free outside Llandudno's Grand Hotel or try the Happy Valley pay-and-display car park by the pier. If you want to pick up the route from **4**, there is pay-and-display parking at the summit of the Great Orme or you can take the tram or cable car up.

START

The walk starts and finishes by the entrance to the pier, near the Grand Hotel, on the seafront in Llandudno.

CONTACT DETAILS

Llandudno Tourist Information Centre, The Library, Mostyn Street, Llandudno, Clwyd LL30 2RP
t: **01492 876413**
e: **llandudnotic@conwy.gov.uk**
w: **greatorme.org.uk**

Ordnance Survey Explorer Map number OL17
© Crown Copyright 2008

Maritime Merseyside
Gateway to the New World

EASY

ACCESS

1½ MILES

2:00

Once one of the richest cities in England, Liverpool has a captivating past and has fully played its part as one of the country's most important ports.

The walk starts outside the Merseyside Maritime Museum at **Albert Dock**.

The city's first dock was built in 1635, remnants of which have recently been uncovered during construction work for the Paradise Street redevelopment. As ships became larger so more modern docks were needed. Albert Dock was opened in 1846. Designed by Jesse Hartley and named after Prince Albert, this was the world's first enclosed dock system. The docks became the lifeblood of Liverpool and the expansion of trade with America saw the city booming. An inscription under the statue of Christopher Columbus in Sefton Park reads, 'The discoverer of America was the maker of Liverpool'.

By the 1960s Liverpool's docks were declining and by the late 1970s the Albert Dock was abandoned and derelict. The newly formed Merseyside Development Corporation took control of the Albert Dock complex in the 1980s and, with considerable government funding, set about transforming the old buildings, which now house shops, cafes, the Maritime Museum and a northern outpost of the Tate Gallery.

The Piermaster's House stands close to the dock entrance to the river. Looking

out across the river from here you can see **Birkenhead**, from where the Mersey ferries ply their trade. To the left of the ferry terminal you can see the steeple of Birkenhead Priory Chapel. This is the oldest building on the Wirral and it was from here that the first Mersey ferries came in 1125.

The distinctive murky brown colour of the river is not due to pollution, it's caused by silt kicked up by the current. Once heavily polluted, wildlife is returning to the river in increasing numbers. One of the many birds to be found today is the cormorant, believed to be the species on which the Liver Bird is modelled.

But the murky water reflects a murky aspect of the city's past. Liverpool's connections with the slave trade are well known. The trade formed a triangle with goods from the UK being shipped out to Africa where they were traded in return for slaves. The slaves were transported across the Atlantic to the West Indies and from there sugar was brought back to Liverpool.

③ A short walk along the riverside will bring you to the **Pier Head**, where there are several reminders of Liverpool's past. During the Second World War the port handled over 75 million tons of cargo and almost five million troops. It was a traffic nightmare for harbourmasters as mixed convoys of warships and merchant ships had to assemble while the bombs fell overhead.

Between July 1940 and January 1942 the Luftwaffe made 68 visits to the city, more than for any other port outside London, and enormous damage was caused, with 4000 people killed, another 4000 seriously injured and 10,000 homes completely destroyed. The worst bombing occurred during the May Blitz of 1941.

Looking across the river to the left you can see the buildings of the Cammel Laird shipyard. During the war the shipyard was

busy turning out ships and submarines at the rate of about one every 20 days. Famous warships were built in the yard, including HMS *Rodney* and HMS *Prince of Wales*.

④ The great buildings that dominate the waterfront, known as the **Three Graces**, are evidence of the wealth generated by Liverpool's position as the second city of the Empire. The middle one is the Cunard Building, built between 1914 and 1918 and once the centre of Britain's cruise ship industry. It was owned by the American Samuel Cunard, whose company later merged with White Star, owners of the *Titanic*, the *Mauretania*, the *Queen Elizabeth* and the *Queen Mary*.

Cruise liners were once a common sight on the River Mersey. Before the advent of air travel, the only way to cross the 3140 miles of ocean to New York was by ship. The liners provided employment for armies of people: painters, cleaners and plumbers would be needed to prepare the vessels for their next voyage. Local laundries would provide the thousands of clean sheets, towels and napkins

JOHNNY WALKER

Near the Pier Head is a statue of Captain Johnny Walker, the greatest hero of the Battle of the Atlantic. Based at Gladstone Dock, his 2nd Support Group hunted U-Boats out in the Atlantic and turned the battle in Britain's favour. Walker sank more U-Boats than any other Allied commander. He died on 9 July 1944 aged 48. After a procession through the streets of Liverpool, Johnny Walker was embarked on HMS *Hesperus* and buried in the waters of Liverpool Bay.

6

COFFIN SHIPS

From January to June 1847, 300,000 Irish immigrants arrived in Liverpool on vessels that were termed 'coffin' ships. Passengers were packed together on deck in all weathers and some ships arrived with a third of the passengers dead.

THE BALTIC FLEET PUB

Opposite the Wapping Dock building stands the Baltic Fleet pub, one of the few surviving dockside pubs. Underneath the pub there are caverns and cellars, and stories abound of a tunnel running to the docks. According to legend, press gangs would get men drunk and then take them through the tunnels and onto waiting ships. They would wake up the following morning with sore heads and the prospect of a long, hard sea voyage.

needed for a voyage and in the 1900s there were approximately 300 laundries in the city.

To your left is the Royal Liver Building, the head office of the Royal Liver Friendly Society. The Liver Birds on top of the building are the largest in the city at 18 feet high with a wingspan of 12 feet. The third building on the right is the Port of Liverpool building, built for Mersey Docks and Harbour Company. This is where all Liverpool ships had to register.

5 In front of you as you look at the water is the area from where **millions emigrated** to a new life in the New World. Walk ahead, through the black gates, and you're taking the same steps as one of those emigrants heading for the ships.

Today ships berthing here mainly carry passengers to and from Ireland, but 150 years ago this area was crowded with immigrants fleeing the terrible conditions caused by the Potato Famine. Two million people, approximately a quarter of Ireland's population, came to Liverpool in one decade. The trip across the Irish Sea on board horribly crowded boats could take three days if the weather was bad.

Many were on route to America. The onward journey was even worse and one in six emigrants who made the journey in 1847 died. Nine million people emigrated to America through Liverpool and at the peak of this tidal wave of humanity a thousand ships a year were leaving the port. The crossing to America could take anything from

four to 14 weeks in berths as small as six feet square for four people. The food and water were unhealthy and disease was prevalent.

Many others stayed and the burden on the city of feeding and housing the Irish immigrants was immense. Lodging houses, mainly in the Vauxhall and Scotland Road areas, were full and the unsanitary conditions caused a typhus epidemic.

6 Turn around and head inland up Water Street and you will see the **Town Hall**. The splendour of this building is a symbol of Liverpool's success, but it has also been a focus for discontent.

The story of Liverpool is one of great wealth and great poverty. Many of the grandest buildings on nearby Castle Street were banks, built to handle money made from the slave trade, but those working the ships saw very little of that wealth.

The Town Hall, built in 1754, is one of the oldest buildings in the city centre and has witnessed many dramatic events, including the seamen's strike in 1775 when police opened fire on a crowd, killing at least two people, and the May Blitz of 1941. But its balcony has also been used for celebrations by the Beatles, members of the Royal Family and both of the city's football teams.

7 Walk down Castle Street and turn right into Derby Square, where you will see a **statue of Queen Victoria**. Some time between 1232 and 1247 Liverpool Castle was built here.

During the Civil War the castle witnessed some of the most dramatic scenes in the city's history. In May 1643 a siege saw Royalist Prince Rupert take the town back from the Parliamentarians and with it the crucial supply route to Ireland. The Parliamentarians, led by John Moore, regained possession of the castle

At first glance it may not appear that there's much here apart from a huge construction site, but these docks have a unique ecosystem without which there would be no businesses here as the docks would fill with algae and stink. Deep in the depths here there are living creatures filtering the water, and the dock is home to jellyfish, mussels, sponges and seaweed. Many of the creatures have come from other countries, transported in the ballast tanks of ships. Giant snails and Korean sea squirts can be found. They were originally brought over in the 1950s and now dominate the **dock walls**. There are also barnacles that have come from as far away as New Zealand and Australia.

Turn around and walk back towards Canning Dock. Bear left and you will be back at the starting point.

the following year and it was left in ruins. It was demolished in the early 1700s and the bricks recycled for use in other buildings.

⑧ Continue along James Street, bear left at the end and walk on along Wapping, bearing right when you get to the Police Headquarters. On your left is Salthouse Dock and on your right is Albert Dock.

Maritime Merseyside

ADVICE
The whole walk is on pavement and tarmac, and is accessible for wheelchairs and parents with buggies.

PARKING
Pay-and-display parking is available at the King's Dock, which is adjacent to the Maritime Museum.

START
The walk starts at the Merseyside Maritime Museum, which is situated in Albert Dock. Paradise Street Bus Station and James Street rail station are a short walk away.

CONTACT DETAILS
08 Place (City Centre Tourist Information), Whitechapel, Liverpool L1 6DZ
t: **0151 233 2008**
e: **08place@liverpool.gov.uk**
w: **visitliverpool.com**

Ordnance Survey Explorer Map number 275
© Crown Copyright 2008

Whitehaven
Undiscovered Georgian jewel

EASY

ACCESS

2 MILES

2:00

DOWN THE PIT

In 300 years over 70 pits were sunk in the Whitehaven and district area. During this period some 500 people were killed in pit disasters. The largest disaster was in 1910, at Wellington Pit, where 136 miners lost their lives. In 1947, at William Pit, there was another disaster of similar proportions where 104 men were killed. Today there is no mining carried out in Whitehaven. The last pit to operate, Haig, was closed in 1986.

Whitehaven was once one of Britain's most important ports. Now its restored harbour offers a great insight into the town's historic marine prosperity.

At the north end of Millennium Promenade you will find a sculpture depicting the fish that brought Whitehaven its first industry. This sculpture, called *The Whiting Shoal*, is the start of the Whitehaven coastal walk.

Fishing has always been important for the town. But its heyday was in Georgian times when shipbuilding and coalmining brought wealth and prosperity, and Whitehaven became the second most important port in England after London.

Today, though still active, Whitehaven harbour is used as much for leisure as for trade. The first dock you come to as you walk

out towards the sea is Queen's Dock. It was here that vessels unloaded their cargo. Now Queen's Dock has become Queen's Marina and is full of leisure craft.

At the back of the marina you can see the remnants of the old sea lock. This was used to allow ships into the dock on the tide. Once the lock was closed the ships could safely be unloaded and loaded up again, ready to sail on the incoming tide.

Today the main harbour on the left is home to Whitehaven's 200-strong fishing fleet. Keeping the harbour on your left, continue past the ice plant and you'll come to the modern sea lock. The gateway to the Irish Sea, the lock is manned 24 hours a day and lets boats in and out independently of the tides. The immense doors keep the inner harbour full of water and act as the front line of the town's flood defences.

2 Having crossed the sea lock you soon come to the Old Quay. In 1778 this was the scene of the only unfriendly American invasion of mainland Britain.

John Paul Jones, a Scot who had learnt his seafaring trade in Whitehaven, left these shores for America. His naval career flourished and he established a reputation as a 'dandy skipper'. However, he fled to Virginia after being arrested for murder following an incident in Tobago.

During the War of Independence John Paul Jones' experience at sea saw him commissioned as first lieutenant in the American navy. In 1778 he set sail for Europe in order to harass British shipping. Presumably because of his connections, Jones decided to mount a hit-and-run raid on Whitehaven. Two raiding parties landed at night and overpowered the pier master. The wooden ships, moored in their hundreds, would have burned well. But once ashore the American sailors got a sniff of the local rum

and beer, and ended up drunk, returning to their ships in disgrace.

The incident is remembered through a series of statues on the South Harbour quayside, called the *Whitehaven Battery*, and the story is told in detail in the **Beacon Centre** on West Strand.

3 Keep walking straight along East Strand to the **market place**. Whitehaven has been a market town since 1654. The old market hall building, which opened in 1881, is now the Tourist Information Centre. Upstairs, close to the site of the original Whitehaven pottery, is a coffee shop run by the new Whitehaven Pottery, a trust that combines educational, tourist and retail opportunities for young people in the town. With the dramatic growth of population in the late 17th century the demand for pottery increased. Aaron Wedgwood, of the famous Staffordshire pottery-making family, visited the town in 1698 to do some trials on the local clay and concluded that it would be suitable for red-ware, stone-ware and a form of semi-china, and with this the Whitehaven pottery industry went into high production.

Pottery manufacturing flourished in the town for 200 years, exporting pieces all over the world. Resurrected in 2003, many of the modern pieces are decorated to commemorate Whitehaven's maritime heritage.

4 From the market place, turn left into Queen Street, past many of the town's Georgian buildings. The oldest of the larger houses is **151 Queen Street**, which was built in the 1730s by tobacco importer William Gale. This double-fronted house is typical of the period, with large reception rooms on the ground floor, many of which have original panelling. The house now stands empty, but it is hoped it will soon be opened to the public.

William Gale's brother George married Mildred Washington during one of his visits to Virginia. Mildred, a widow, had three children from her first marriage: John, Augustine and Mildred. They returned to Whitehaven in 1699. Mildred died in childbirth soon afterwards.

DODGY BUSINESS

Captain John Paul took command of the *Betsy*, a West Indian ship, and remained for some time in the West Indies in commercial business. He seems to have accumulated considerable wealth. In 1773, however, he had to leave the West Indies after he killed the ringleader of a mutiny in a dispute over wages. Local feeling was against him and he fled to Virginia, changing his name, first to John Jones and later to John Paul Jones.

IT'S A WONDERFUL TOWN

Whitehaven is the most complete example of planned Georgian architecture in Europe. Due to its rigid layout, with streets running parallel or intersecting at the perpendicular, many historians believe that it was the blueprint for the New York grid system.

After her death, the executors of her first husband's estate contested Mildred's will and the two remaining children were sent back to Virginia. As an adult, Augustine fathered George, a boy who became President George Washington in 1789. The square at the front of 151 Queen Street has been named Washington Square in honour of the town's presidential connection.

5 Walking on from 151 Queen Street, turn right onto Roper Street. The **Georgian architecture** continues. Many of these grand houses belonged to some of Whitehaven's more influential people.

Daniel Brocklebank, shipbuilder, lived at number 25. As well as building ships for others, in 1795 he boasted his own fleet of 11 vessels. The company evolved into the shipping line T & J Brocklebank and went on to become the Cunard Line.

Further along, number 30 belonged to James Spedding, son of Carlisle Spedding. Carlisle Spedding was a mining engineer who, together with the Lowther family, helped bring business and wealth to Whitehaven. Spedding invented the steel mill, a safety device that detected if gas was present in a mineshaft.

6 At the end of Roper Street turn left up Scotch Street and left again into Lowther Street, one of the broadest streets in the town's grid system. Halfway along Lowther

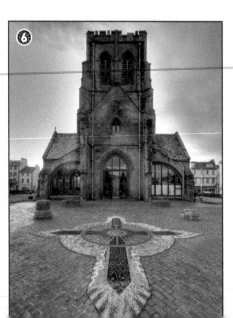

Street stands **St Nicholas's Tower**. This tower is all that remains of St Nicholas's Church, built in the 1880s. One August afternoon in 1971, a fire destroyed the nave and sanctuary, leaving only the tower and the side chapel. Mildred Gale, the grandmother of the first American president George Washington, is buried in the churchyard.

7 Continuing along Lowther Street, towards the harbour, you'll find the **Rum Story**, a museum housed in the original offices of wine merchants Jeffersons. A partnership with Lumley Kennedy and Co Shipbuilders enabled Jeffersons to import sugar from their estate in Antigua, rum and molasses from the West Indies, and wine from Portugal and Spain.

In the 1600s Whitehaven had been the leading importer of tobacco from Virginia. Following defeat in the American War of Independence many of the Virginia settlers moved to the Caribbean to continue their trade. Once there they expanded, growing tobacco and sugar cane, and manufacturing rum. The museum tells the story of the making of this sweet liquor. Of course, the rum industry was based on another altogether more sinister one. In common with many other ports up and down the country Whitehaven was involved in the slave trade.

8 For the last stretch of the walk, turn left and head back towards the harbour, down Lowther Street and across Strand Street. Make your way out onto the Lime Tongue, at the end of which sits the **Crow's Nest**. The

Lime Tongue was the third of the harbour extensions, built in 1754 at the height of the town's prosperity. It is believed that the name comes not from the trade with lime fruits, but from the lime used in the production of cement. To your right is the Sugar Tongue. Built in 1735, this was the second quay, built to appease the merchants who were calling out for more space to unload their cargoes.

In the early 19th century trade in Whitehaven began to decline as the shallow waters of the Solway couldn't cope with bigger vessels and ports with deeper waters like Liverpool, Glasgow and Bristol prospered. With your back to the sea, you can see a number of buildings that relate to the harbour's past. One of the oldest is a tobacco warehouse, which dates from the 1600s. Now converted into offices, the building stands as testament to the town's heritage as a mighty port.

To return to the starting point, turn round and walk back to shore. Turn left and walk along Millennium Promenade.

MILLER'S MERCY

For ten years slaves arrived at the quayside from the Caribbean, until the movement of the abolition of slavery began around 1769. A Whitehaven tanner, William Miller (1816–1856), was well known for organising meetings to encourage people to stop this awful trade.

Whitehaven

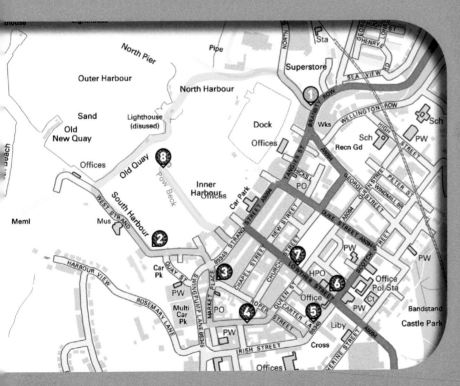

ADVICE
The walk takes you over cobbled streets and on some uneven surfaces. Take care when walking in the harbour as it might be slippery and the old quays aren't always buggy- or wheelchair-friendly.

PARKING
There are numerous pay-and-display car parks close to Whitehaven harbour. Queen's Dock car park is near to the start, while Quay Street South car park is closer to the end at Crow's Nest on Lime Tongue.

START
The start of the walk is at *The Whiting Shoal* sculpture between Tesco's car park and the Millennium Promenade.

CONTACT DETAILS
Whitehaven Tourist Information Centre, Market Hall, Market Place, Cumbria CA28 7JG
t: 01946 852939
e: thebeacon@copelandbc.gov.uk
w: visitcumbria.com

Ordnance Survey Explorer Map number 303
© Crown Copyright 2008

EASY

ACCESS

2 MILES

🕐

3:00

Belfast
No mean city

A walk through this exciting city reveals how a sandy ford became a powerhouse of trade and industry, a hotbed of political unrest and the place where Sherlock Holmes bought his overcoats.

CITY OF INVENTION

At the turn of the 20th century Belfast was known as 'Linenopolis', and it led the world not only in the production of linen but of huge ships, ropes, tea-drying machinery and aerated water. It was a city of innovation, inventing, among many other things, milk of magnesia, air-conditioning and barbed wire.

1

Young, brash and with lots of cash to flash, that was Belfast a hundred years ago when the City Hall, where this walk starts, was nearing completion. This confection in Portland stone and Italian marble was intended to show that this dynamic industrial and commercial powerhouse was no mean city.

From when it opened in 1906 the City Hall was to play a central role in life of the city. It was the setting for the Signing of the Covenant in 1912, the seat of the first parliament of Northern Ireland after partition in 1921, the focus of huge rallies and demonstration such as the VE Day celebrations in 1945, and

homecoming welcomes for Olympic gold medallist Mary Peters and world boxing champion Barry McGuigan visited and United States President Bill Clinton.

Conan Doyle when he dressed his fictional detective Sherlock Holmes in one.

3 Further down High Street is **St George's Church**, the spot where settlement began and which gave the city its name. There has been a church here for more than a thousand years, initially so that travellers could give thanks for, or pray to have, a safe crossing of Beal Feirste, the sandy ford at the mouth of the Farset River. The name Belfast is the anglicisation of Beal Feirste.

In the 17th century, Protestant settlers from Scotland, England, and some from the Isle of Man, were moved in and a town began to grow. Before long the banks of the Farset became the first quaysides of a merchant city. What was the Farset River is now High Street and a tunnel big enough to take a bus now carries the river underneath.

4 Carry on down High Street until you get to the river. This **quayside** used to be one of the busiest in the city, where freight came and went, where thousands of emigrants began their journey to new lives overseas and to where cattle, pigs and sheep were driven from the markets in Oxford Street, to be carried away to the slaughterhouses of Great Britain.

From the quay you can also see the giant cranes, Samson and Goliath, which tower over the biggest building dock in the world in the Harland & Wolff shipyard. The yard has shrunk dramatically since its heyday (its workforce numbered up to 30,000 in the past) and is now devoted to design and repair. There are slipways up to the left, and beyond the bridge, as you look across the river. But if you'd been able to look across in 1912 you'd have seen the *Titanic* being launched from one of those slipways and undergoing her sea trials in Belfast Lough.

2 Turn right outside City Hall, then left up Arthur Street and Cornmarket. On the **corner of Cornmarket and High Street** is the spot where one of Belfast's leading citizens and rebel leader, Henry Joy McCracken, was hanged on 17 July 1798.

McCracken had led a band of rebels in the battle of Antrim and was hunted down after the rebels were defeated. An offer was made to spare his life if he named his co-conspirators, but he declined and sealed his fate. The gallows were erected outside what is now Dunnes Stores, where the town's market house stood in 1798.

The market house is long gone, but commerce remains. High Street became, and remained for a long time, the city's prime shopping area. It has been undergoing a steady recovery from years of decline and difficulty, especially during the years of the recent Troubles.

One notable former business in High Street was the Ulster Overcoat Company, which manufactured a coat known as the Ulster. This was made famous by Arthur

HARLAND & WOLFF

Belfast was the natural place to build such an ambitious and headline-grabbing ship as the *Titanic*. Harland & Wolff had been building on a grand scale for years by 1912. In 1899 the *Oceanic*, the largest ship built anywhere in the 19th century, slipped into the water and the record books here. It was overtaken in 1901 when the *Celtic*, the largest man-made moving object ever built in the world until then, was launched. But it wasn't just the ships that set records. In 1918, in one nine-hour shift, James Moir drove 11,209 red-hot rivets into the plates of a warship being built in H&W, a record for a working shift that was never equalled nor surpassed.

DUBARRY'S HOUSE OF ILL REPUTE

Across from the Customs House stands McHugh's pub and restaurant, one of the oldest surviving buildings in Belfast, which has recently been renovated and extended. The extension took in and brought an end to one of Belfast's most notorious pubs, Dubarry's, which was once known to seamen across the world as Belfast waterfront's busiest house of ill-repute.

SCOOP

The award for the greatest scoop published in Belfast has to go to the *News Letter*, whose enterprising publishers managed to intercept the American Declaration of Independence on its way to London from Philadelphia in 1776 and print its contents before it reached its destination.

5 Cross the river on the **Queen's Bridge**. When the bridge was officially opened in 1849 by Queen Victoria, Belfast had embraced industrialisation with enthusiasm and was expanding rapidly. The bridge was a vital link between the old town and its quays, and Ballymacarrett, on the County Down side, which was already well on its way to becoming the industrial heartland of the city.

Turn right and walk along Queen's Quay. This side of the river was to be home to the world's biggest ropeworks, shipyard, aerated water manufacturer and manufacturer of tea-drying machinery, among other things. The linen mills, shipyards, engineering works, foundries and factories all needed labour and Catholics and Protestants poured in from rural areas, bringing all their political and religious animosities and grievances with them.

These differences had already exploded into violence in the overcrowded poverty of the streets of hastily built houses and a pattern had been set for generations. When rural migrants came to Belfast they settled in areas already occupied by fellow Catholics or Protestants and the layout of working-class neighbourhoods remains broadly the same today.

When Queen Elizabeth II opened the Queen Elizabeth Bridge in 1967 she was reminded of the tensions that had never been far from the surface since her ancestor had opened the neighbouring bridge more than 100 years before: her car was struck by a concrete block as it drove along Great Victoria Street.

6 When you cross back over the river, turn right and head up to **Customs House Square**. The Customs House itself is an imposing Victorian building, put up as Belfast became one of the great industrial and trading centres of the Victorian United Kingdom.

In the 19th and early 20th centuries, the steps outside the Customs House became the Speakers Corner of Belfast, where orators harangued, exhorted and cajoled the crowds enjoying the open space of the square. This is where dock labour organiser Jim Larkin held court. Larkin was sent to Belfast, from Liverpool, in January 1907 by the National Union of Dock Labourers to organise dockers in the city.

In May that year a dispute with one employer escalated and soon the city was paralysed by strike and lockout. Protestant and Catholic dockers were united in their support for Larkin, as were the carters who hauled the goods to and from the docks.

Attempts to curb the pickets were foiled when men of the Royal Irish Constabulary mutinied. An attempt by the employers to create a Protestant workers-only union also failed and the dispute was eventually settled with a grudging recognition of the dockers' union and a pay rise for the carters.

7 Turn left out of Albert Square and walk along Waring Street. Where Bridge Street and Waring Street meet you can look across to an empty building. This was the site of **an exchange and assembly rooms**, which in the 18th century was the heart of Belfast. The building was commandeered for a military tribunal in 1798 in the wake of the failed rebellion by the United Irishmen and it was here that Henry Joy McCracken was tried and sentenced to death.

McCracken was a resident of Rosemary Street (to the left as you look at the assembly

building) and publisher of the Belfast *News Letter*. Like other reformers and radicals McCracken was from a family made prosperous by Belfast's success as a trading port, but he felt oppressed by English rule, and Anglican religious authority and tithes.

There are only a few buildings of any age in this part of Belfast because the area was devastated in a German air raid on the nights of 4 and 5 May 1941 and most of the buildings were reduced to rubble.

8 Walk along Donegall Street and into Writers' Square. This open space is a key element in the Cathedral quarter's reinvention as a centre for the arts and, when skateboarders aren't rattling over the quotations about Belfast by famous writers inscribed underfoot, it's a venue for street performers and arts festival events.

Directly opposite the square is **St Anne's Cathedral**. It is built mainly of Portland stone, but a stone from every county in Ireland was used in its construction. Work on the cathedral took more than 80 years and seven architects to complete.

The streets behind the cathedral were once part of the city's Sailortown docks area and were known as the Half Bap. Most of Sailortown was swept away for redevelopment, but some of the character can still be glimpsed in the streets.

From here you can walk back along Royal Avenue and Donegall Place to the starting point outside City Hall.

Belfast

ADVICE

The whole walk is on pavement and tarmac, and is easily accessible for wheelchair users and parents with buggies. There is an incline onto the footbridge over the Lagan at Donegall Quay, but the ramp is in a spiral and not too steep.

PARKING

There are a number of pay-and-display car parks within walking distance of City Hall, including Grosvenor Road and Chichester Street.

START

This is a city centre route beginning at City Hall and ending a short distance away from there in Donegall Street.

CONTACT DETAILS

**Belfast Welcome Centre,
47 Donegall Place, Belfast BT1 5AD**
t: **02890 246609**
f: **02890 312424**
e: **welcomecentre@belfastvisitor.com**
w: **gotobelfast.com**

Permit Number 70230, Ordnance Survey of Northern Ireland
© Crown copyright 2008

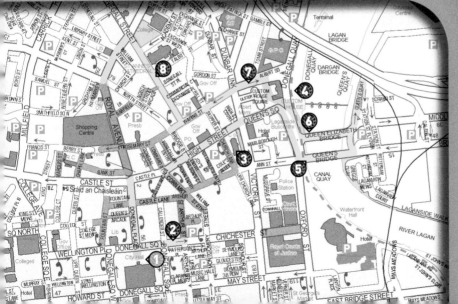

The Causeway Coast
The dual delights of Antrim

HARD

ACCESS

12 MILES

5:00

This section of the Antrim Coastal Path links two of Northern Ireland's best-known visitor attractions: the Giant's Causeway and the Carrick-a-Rede rope bridge.

1 Start at the Giant's Causeway Visitor Centre and take the main route down to the causeway. On the right as you go down is an outcrop of red laterite, an iron-rich rock which appears frequently along the coast here. On the left, on the other side of the bay, is the Camel's Back. The causeway was developed as a tourist attraction in the 19th century and many of the rock formations acquired romantic names – often rather tenuously.

You reach sea-level at the causeway itself. Follow the path through a large cleft in the rocks and start heading uphill. At the junction keep on along the coast and follow the old lower path as far as you can. It is now closed because of rockfalls and unstable cliffs, but gives some excellent views of more romantically named formations – **the Organ** (vertical shafts), the Harp (curved shafts) and the Amphitheatre (just what you would expect).

Although the causeway is the most famous section, the hexagonal basalt columns extend

much further and form the cliffs along which you will be walking for several miles. The distinctive shape arose during the cooling of lava flows. As the material cooled, stresses built up inside and the stored energy eventually became enough to form new surfaces, in much the same way that the stored energy in a toughened windscreen causes it to form small regular pieces when broken.

2 From the gate at the end of the path, turn back to the junction and this time head uphill on the Shepherd's Steps. When you get to the top, have a rest (there are 162 steps) and then turn left along the coastal path. You will be following this all the way to Carrick-a-Rede and it is reasonably well signed all the way. In good weather the views from the cliff-tops are spectacular. To the west is the entrance to **Lough Foyle** and beyond the mountains of Donegal. To the north is Scotland. The most westerly island is Colonsay, then Islay (which appears as two separate pieces – Loch Indaal is in between) with the hills of Jura beyond. Further east is the Mull of Kintyre, with the low valley between Campbeltown and Machrihanish visible. The large island much nearer is Rathlin.

3 After Contham Head the path crosses a stile, passes the remains of a wartime watch post and then starts descending steadily to Dunseverick Castle. Shortly after the bay of **Port Moon** (look for the cottage at the bottom, which is a former salmon fishery, and its amazing steep access track) the path

crosses another stile and runs inside the cliff-top fence for the first time. Skirt the edge of this field, go through the kissing gate and turn right past the edge of the picnic area.

4 **Dunseverick Castle** was built by the MacDonnell clan in the 15th century and in 1642 was destroyed by a Scottish army led by General Munro, while suppressing a rebellion led by Rory O'More, Lord Maquire and Sir Pheilim O'Neill. Four hundred and fifty years of decay have taken their toll and there is little left to see now.

As this point, if you have had enough already, the Causeway Rambler bus (summer only) will take you back to Giant's Causeway. If you want to go on, but don't fancy scrambling along the shore over rocks and boulders, you can take an alternative signed route along the road to White Park Bay.

Otherwise walk along the road for a short distance (50 yards) before turning back onto the path at the stile by the telephone pole. The path goes steadily downhill and over a wooden bridge – now bear left to the shore, round the head of a small, rocky bay and then over a couple of headlands until you reach a small road leading to Dunseverick Harbour. The harbour is owned by the National Trust, and has a car park and toilets.

5 Leave the harbour by taking the steps which climb the side wall beside the mounted anchor and then crossing a stile. Go round the head of the beach, and make for the wooden steps and handrail at the far side, then follow the path as it winds its way along the shore

THE *GIRONA*

The *Girona* was part of the Spanish Armada in 1588. She was wrecked off Lacada Point, just east of Giant's Causeway, in October 1588 with the loss of all but nine of the 1300 on board. She was carrying valuables from two other wrecked ships of the Armada as well as her own. Some guns and valuables were salvaged at the time, but much more treasure was recovered in 1967 and 1968 and is now in the Ulster Museum.

RATHLIN ISLAND

Rathlin is only 11 miles from the Mull of Kintyre, and its ownership was disputed between Scotland and Ireland for many years. In 1617, so the story goes, the matter was settled by taking a snake to the island. The unfortunate creature died, so Rathlin was deemed to be Irish – as St Patrick had expelled all snakes from Ireland.

SALMON FISHING

The Antrim coast once supported a series of salmon fisheries, using purse nets. One end of the net was attached to the shore and the other was anchored some distance out. Salmon were swept into the net by the tide and emptied into a boat working its way along the net. The net at Carrick-a-Rede island was the shortest on the coast, at 230 feet, and the rope bridge was built for access by the fishermen.

and low cliffs. Shortly after the path goes through a rock arch you pass a disused salmon fishery and arrive at Portbraddon. The tiny **church of St Gobhan** is just off the road on the right-hand side – it is claimed to be the smallest church in Ireland. At the far side of the bay leave the road to walk round the shore

to White Park Bay. This section involves some energetic scrambling over large boulders and if you prefer to avoid this you can take the longer, but flatter, route by road instead.

6 The cliffs are now chalk with a lot of embedded flint. Chalk is relatively soft and easily eroded, so you need to climb over large fallen lumps until you reach an impressive cave in the cliffs. Beyond the cave the going gets much easier, with flat beds of chalk and then the smooth sand of **White Park Bay**. Walk to the far end of the beach, as near to the sea as you dare for firmer going. At the far end don't climb the hill, even though it looks inviting with obvious paths. The path continues at sea level round the low cliffs at the extreme east end of the beach. It is very hard to spot from any distance, but obvious when you get there.

7 The path leads to **Ballintoy Harbour** along a rugged section of coast with arches, inlets and fjords – reminders of the force of Atlantic storms. The harbour was developed to ship out granite setts produced at the nearby Larry Bane quarries and is still a working harbour for local fishing boats. There is a picnic and barbeque area on top of the limekiln; car parking, toilets and a tearoom in one of the harbour buildings.

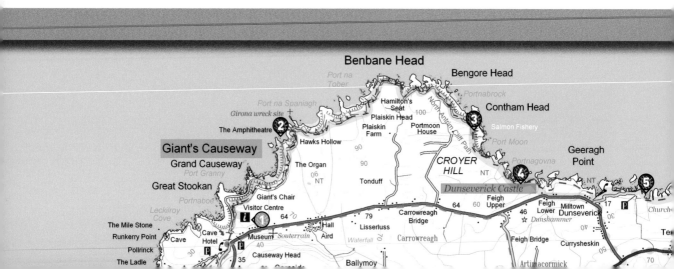

For the leg-weary who really can't go on to the end of the walk the Causeway Rambler bus stops in Ballintoy village, about ten minutes' walk from the harbour.

The island just offshore here is Sheep Island, where a few sheep used to spend the summer. Not too many though – the local wisdom was that Sheep Island would fatten 'ten sheep, feed 11 and starve 12.

⑧ For the last section, to Carrick-a-Rede, take the road up and away from Ballintoy

Harbour and to the church. Where the road turns sharply right after the churchyard turn left instead, onto the fenced-off footpath. After only half a mile or so pass the former quarries on your left and arrive at the car park for **Carrick-a-Rede rope bridge**. The bridge is owned by the National Trust and is reached by a path starting at the top of the car park. Allow a minimum of 30 minutes to walk to the bridge, cross it both ways and return. It was originally built every year by salmon fishermen for access to the island, and is still taken down each autumn and re-erected the following spring. The drop to sea level is around 85 feet.

The Causeway Rambler bus back to the Giant's Causeway starts from the car park here.

The Causeway Coast

ADVICE

As well as the Shepherd's Steps this walk includes numerous shorter flights of steps, frequent climbs and descents, a scramble over large chunks of fallen chalk on the beach at Portbradden and over a mile along sand at White Park Bay. There are toilets, car parks and tearooms at a number of points along the walk. The Causeway Rambler bus runs hourly through the day from early July until late September. Carrick-a-Rede rope bridge is open from the start of March until the end of October.

PARKING

Park either at the Giant's Causeway Visitor Centre or in the railway station car park. You can also get there by bus and train from Coleraine.

START

The walk starts at the Giant's Causeway car park.

CONTACT DETAILS

Giant's Causeway Visitor Centre, 44 Causeway Road, Bushmills, County Antrim, Northern Ireland BT57 8SU
t: 028 2073 1855
e: info@giantscausewaycentre.com
w: giantscausewaycentre.com

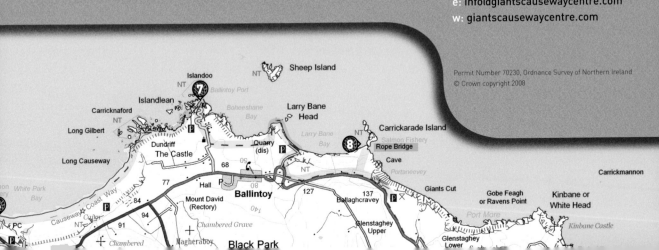

Permit Number 70230, Ordnance Survey of Northern Ireland © Crown copyright 2008

Portpatrick
The Scottish–Irish connection

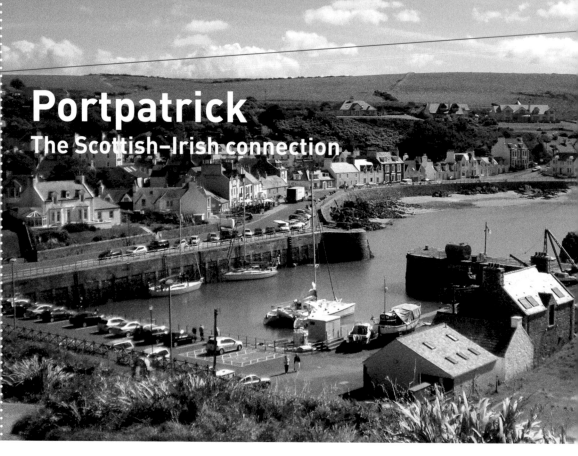

HARD

ACCESS

8 MILES

3:00

Northern Ireland at Donaghadee is only 22 miles away from Scotland at Portpatrick. This walk explores the fascinating consequences of this closeness.

1

Start in the car park by the pier in Portpatrick. The inner section of the pier was built by John Smeaton in 1774 while the outer section is part of the remains of a grand harbour designed and built by John Rennie, starting in 1821. Works had not finished when a terrible storm of 6 January 1839 did a huge amount of damage and the harbour was never completed. A lighthouse had been built in 1836 at the far end of the south pier, but it was undermined by the storm and replaced by the one you can see now. The original lighthouse was dismantled and reassembled in Colombo in Sri Lanka, where it can still be seen.

Walk north from the lighthouse along the promenade, which was built to carry a plateway (tramway) for carrying stone for Rennie's North Pier. On the land side is a

1

lime kiln, built into the cliff face at the head of the raised beach. This was used to produce mortar for the harbour works. Shortly after this is Barrack Street: Portpatrick was an important garrison town and staging post for troops travelling to or from Ireland.

2 Turn right up Main Street. This was laid out about 1800 when a straggle of pre-Reformation monastic buildings – by then used as military accommodation – was cleared away. Soon afterwards turn left into St Patrick Street. The large white house in front of you is Inglenook and was built as the manse (minister's house) in 1726. Portpatrick had a thriving business in marriages for Irish runaways – it fulfilled much the same role as Gretna Green did for English couples.

The remains of the church may be entered by the gateway immediately next to Inglewood. The church itself was built in 1622–1629 and was based on the remains of a much earlier church that had fallen out of use after the reformation the century before. The most striking feature is the **circular tower**. Its origins are unknown. One theory suggests it is Celtic, while another posits that

it served as a lighthouse (or leading light) for the harbour. The churchyard has a splendid collection of gravestones, particularly of sailors, with fulsome epitaphs, many of them in verse.

3 Return to Main Street and turn left. Ahead of you are the remains of the bridge that carried the railway to Portpatrick Harbour station. Turn left down Dinvin Street, cross the bridge and bear left to follow the road beside the burn. The railway ran behind the cottages on your right to a station where the tennis courts and bowling green are now. Portpatrick Harbour station has the distinction of being one of the shortest-lived stations ever built: it opened to traffic in September 1868 and closed around two months later. The branch line from the main station was very steep (1 in 35) and could only be used by trains of four wagons – and expected traffic from the harbour never materialised.

Walk round the enclosed **inner harbour**, where the remains of Rennie's North Pier can be seen extending out from the north face of the small outer section. His original design was for a U-shaped harbour with McCook's Craig as an island in the middle. There has been a lifeboat station here since 1877. The Portpatrick lifeboat was involved in the *Princess Victoria* disaster of 1953 and a monument to that rescue can be see on the cliff face beside the Dinvin Burn.

4 The town section of this walk is over and the cliff section to Killantringan starts at the steps beside the children's playground. This is the start of the **Southern Upland Way**, which runs 212 miles to Cockburnspath on the east coast and is well marked all the way. A historical timeline is cast into the concrete at the foot of the steps and geological information is cast into the steps themselves. Shortly after the top of the steps are the buildings and antenna of Portpatrick Coastal Radio Station (callsign GPK). It started as a part-time naval radio station in 1905 and later became a 'working station', controlled remotely by radio operators in Stonehaven. The very last commercial Morse

IN MEMORIAM

Brother Seaman,
pass not by, my shattr'd
hull lies here
My Anchor's cast, when
next it's weigh'd full rigg'd
I shall appear,
But tho' I'm shipwreck'd
in the tomb, my Hull thus
cast away,
I only wait to know my doom
until the Judgment day.

Epitaph of John, Robert and Andrew McCormack, Portpatrick churchyard

PRINCESS VICTORIA

The *Princess Victoria* was an early roll-on-roll-off car ferry, built in 1947. On 31 January 1953 she sank in a storm while crossing from Stranraer to Larne. Her stern doors were battered open by waves and water flooded the car deck, eventually capsizing her with the loss of 133 lives. The radio operator, David Broadfoot, remained at his post, transmitting in Morse to Portpatrick Radio Station as the ship sank. He was posthumously awarded the George Cross.

UNDERSEA ENGINEERING

As well as water, underwater cables have to be able to resist constant abrasion as tidal currents move them over the seabed. The Victorians devoted much ingenuity to tackling this problem and produced some extremely successful solutions, clearly visible at Port Kale. The northern cable (1854, Whitehead) has wires surrounded by layers of copper, gutta percha, linen tape, hemp and steel, while the southern cable (1853, Donaghadee) is similar but simpler.

code transmissions from Britain were on 31 December 1997 using the Portpatrick callsign – acknowledgment that it was then the longest serving radio station in existence. Nowadays it exists only to broadcast Navtex weather and safety information for ships.

5 After the main radio station site go up the wooden steps and turn left onto the tarmac road, which turns into a track. At the end of the golf course the track descends to sea level at Port Mora, known locally as Sandeel Bay. On the right, just as you reach the shore, are two caves. A burn falls across the mouth of the second one, the Cave of Uchtriemacken, and just below the entrance is the Dipping Well. Until early in the 19th century there was a tradition of bringing sick people, and children with rickets in particular, here on the first night of May, to be washed in the stream and dried in the cave.

6 At the far side of the beach at Port Mora the path climbs up steep steps and then descends into the next bay, Port Kale (Laird's Bay). If you wish to avoid the steps, take the path inland as it's only moderately steep and has no steps. When you get to a junction with a bigger track turn left for Port Kale.

Port Kale was the terminating point for the first two successful **telegraph cables** to Ireland. They were laid by the English and Irish Magnetic Telegraph Company. The first was finished on 23 May 1853 and went to Donaghadee and the second went

to Whitehead in 1854. The cables have long gone, but if you look at the edge of the beach under the marker pole you can see the two ends: the 1853 cable to the south and the larger 1854 cable to the north.

7 The walk from here involves several very steep stone stairways. Cross the stream by the bridge and follow the Southern Upland Way markers north. After the two sets of stairs the path crosses a stile. Although the worn track stays inland, it is worth making the very short detour to the headland with the waymarker on it. In good conditions the view is spectacular. To the south are the Rhins of Galloway, with the Isle of Man beyond. Directly west are the Copeland Islands with Donaghadee – the other end of the proposed Portpatrick ferry service – beyond. Carrickfergus is clearly marked by the power station chimney and the Irish coast is visible. Directly to the north is the Ayrshire coast.

The path is now broad and grassy as it traverses Ouchtriemakain Moor and, as you cross a small rise, Killantringan Lighthouse appears. After the kissing gate a tiny beach is visible at the bottom of a **rocky fjord**. After the Second World War, huge amounts of munitions, including phosphorus incendiary bombs and nerve gases, were packed into surplus vessels and sunk in the Beaufort Dyke, a deep (over 650 feet) trench in the seabed between Scotland and Ireland. The containers are now breaking up and the cargoes are escaping. The poison gases are immediately neutralised by seawater, but phosphorous flares regularly wash ashore and it is not unusual to see piles of seaweed on the beach burning quietly. Avoid them.

8 The headland before the lighthouse at Portamaggie gives a view down into Killantringan Bay where, at low tide, the remains of a ship can be seen. The *Craigantlet*

was an 800-ton merchant ship en route from Liverpool to Belfast when she ran aground on 26 February 1982. She was too badly damaged to be removed and has been breaking up in situ ever since.

Follow the path north until it intersects the track to the lighthouse and turn left. **Killantringan Lighthouse** was one of the 26 built around Scotland by David Alan Stevenson between 1885 and 1937. It was opened in 1900 and closed on 11 July 2007 after a review of light provision concluded that it was no longer necessary.

This walk now returns the same way to Portpatrick.

Portpatrick

ADVICE
Portpatrick has shops, hotels, guesthouses and a lifeboat. There are public toilets at the north end of the inner harbour.

PARKING
There is ample free parking on the quayside at Portpatrick. There are buses between Portpatrick and Stranraer, from where buses and trains run to Glasgow and Carlisle (change in Kilmarnock by train), and ferries run to Belfast.

START
The walk starts in the car park by the pier in Portpatrick.

CONTACT DETAILS
**Stranraer Tourist Information Centre, 28 Harbour Street, Stranraer, Dumfries and Galloway DG9 7RA
t: 01776 702595
f: 01776 889156
e: stranraer@dgtb.visitscotland.com
w: visitdumfriesandgalloway.co.uk**

Ordnance Survey Explorer Map number 309
© Crown Copyright 2008

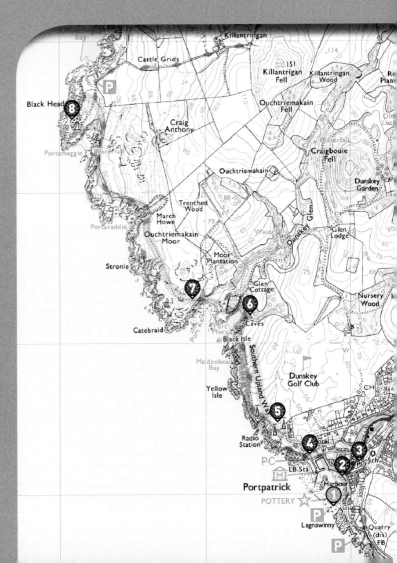

Culzean
Robert Adam's last masterpiece

EASY

ACCESS

2½ MILES

2:00

THE POWDER HOUSE FUNNEL

The design of the Powder House funnel is perhaps the most ingenious part of the story of this unusual building. The round chamber played the part of a funnel, which absorbed the force of the gunshot by pushing the energy upwards into the air.

A romantic castle set in a dramatic cliff-top setting plays host to a walk full of nature and history.

❶ There can't be many houses that have their own private power station, but this walk begins at Culzean's very own **Gas House**.

In the 18th and 19th centuries there was a fashion among landowners for schemes aimed at the 'improvement' of their estates and, while most of these endeavours involved agriculture, at Culzean there was energy left over to take an interest in technology.

One such new fad was the arrival of gas lighting, something that must have seemed like a miracle to those who first witnessed it. The earl himself must have been impressed because he was soon planning to have gas lighting installed. His first problem was how to get the gas in the first place. The solution was found in lumps of a common substance already being mined throughout much of the surrounding county – coal. The Gas House was built, effectively, as a coal-burning power station.

The coal was fired and burned, and the coal-gas was then distilled off from the burning coke. This procedure would never be allowed today as coal-gas is made up of highly flammable hydrogen and methane with small amounts of poisonous carbon monoxide – an extremely dangerous mixture.

It's no surprise then that the Gas House was situated as far away from the actual castle buildings as possible. What is surprising is that anyone volunteered to work there in the first place.

2 Scrambling over the rocky shore to the front of the castle may be a bit arduous, but you will be rewarded when you come to some **caves dug into the rock** under the castle and decorated in the Middle Ages to look like inhabited caves.

In fact, the caves have been here for thousands of years as the castle itself is built on a network of caves that were probably inhabited during the mediaeval period.

After this period, and certainly by the time the present castle buildings were constructed, it seems likely that these caves were used for smuggling. In this secluded and privately owned part of coast it was easy for those bringing contraband goods, such as whisky, tobacco and silks, into the country to unload and store their goods here.

It's hard to believe that, with the high level of smuggling going on, the Kennedy clan who owned Culzean, or at least their agents who ran the castle, were not either heavily involved in it or at least tacitly allowed it to continue for a share of the profits.

3 Nestling on the cliff-top path, enjoying a vantage point looking out over the Firth of Clyde, is an oddly shaped building standing all on its own. This is the **Powder House**. It may seem to have been designed with the object of enjoying the view, or as a quaint Victorian folly, but in fact this building performed a more bizarre function.

At 8 o'clock every morning during the halcyon days of Culzean the estate resounded to an almighty boom, which emanated from this small castellated outpost, the home of the estate gun. Perhaps one of the most extravagant alarm clocks in history, the two-pounder fired a shot to announce the new day.

This tradition continued right up until the First World War, when it was abandoned, quite probably to stop anyone thinking it heralded a German invasion.

4 Further along the cliffs from the Powder House you can walk deeper into the deciduous woodland of Culzean, where some five million trees were planted by Robert Adam during his landscaping of the grounds.

At **Carrick Point** there are more wonderful natural sights to behold. Looking down at the cliffs from this vantage point you can see some of the many hundreds of species of birds that live here. All the usual seabirds are here, as well as some more unexpected species, such as barn owls, fulmars, gannets and eider ducks. Another common sight are heron, which are often to be seen feeding in the rock pools before flying back to sites inland where they rest. The rock pools are densely populated with many varieties of life,

from small wading insects up to the squat lobster. You may also see grey seals basking on the rocks offshore.

5 After stretching your legs around the cliffs, take a few moments to relax and admire some of the woodland wildlife nearby. Formed from the damming of a stream by Robert Adam, **Swan Pond** has become one of the top attractions on the estate. There is a hide at the northern end from where it is possible to observe the fauna without disturbing them.

Alongside the graceful swans, which have bred successfully here, there are several species of duck. There are frogs, toads and the bat house is home to several rare species, including both varieties of pipistrelle. However, the highlight is undoubtedly the resident population of otters.

It takes hard work to keep the pond environment stable to support this range of wildlife. One of the biggest challenges is keeping the water clear of water lilies. The lilies can affect water quality by blocking out the sunlight, causing problems for water-based creatures and thus altering the ecology of the pond. The ingenious solution to this is to employ a machine like an aquatic lawnmower, which disturbs the roots of the lilies, keeping them in check.

6 One of the most beautiful parts of the castle is the walled garden, with its beautiful beds and superb herb garden. It's also well worth taking a look at the fantastic vegetable garden and the fully working vinery.

KEEPING THE VINES WARM

Today the vinery uses a clever system for maintaining the temperature. Animal droppings are placed in the dark-coloured hatches under the vinery. As the black material absorbs the heat of the sun, the dung heats up and releases warm gases, which then flow through pipes keeping the vinery warm.

The walled garden is split into two sections. The north section contains a large working garden, with vegetables and a large selection of fruit trees on the western wall, which turns the corner into an impressive bed of sunflowers of all varieties. The centrepiece is a beautiful double herbaceous border, a haven for butterflies in the summer. The south section contains the herb garden.

Over on the western edge of the garden sits the working vinery. In the 18th and early 19th centuries a **vinery** was one of the ultimate status symbols. Not only did it show you had the considerable wealth needed to build one in the first place, but also that you could afford a highly talented gardener to keep the vines alive.

7 One of the architectural gems of Culzean is the **camellia house**, designed by Robert Adam. Nestling on a low rise between the walled garden and the castle itself, it looks glorious when bathed in morning sunlight. This construction is a spectacular Gothic creation, restored to its former glory around ten years ago. It was originally built as an orangery, but was soon housing camellias, as well as figs and acacias, all staples of the exotic tastes of the Victorian estate.

8 The walk ends in the castle's main courtyard, after you've walked through the 'ruined arch' designed by Adam to give Culzean an antiquated classical feel, but in reality no older than any other part of the castle.

Culzean was the last masterpiece of Robert Adam, the celebrated Scottish architect and designer. In 1777 Adam was commissioned by David Kennedy, the 10th Earl of Cassilis, to completely renovate the old castle and the work was completed in four stages over the next 15 years.

The third stage of this development, in 1785, was the most dramatic, for it was at this point that Adam added the drum tower, which has become Culzean's signature. This elegant structure perched on the edge of the cliffs symbolises Adam's entire style: the classical lines of the castle existing in harmony with the rugged, romantic natural rocks.

If you stand with your back to the wall of the castle, you will see a marvellous panorama open up before you. From Cumbrae and Bute in the north, the view swings round to the west taking in Kintyre and, on a sunny day, Jura. If you're lucky and get a clear day, it's possible to look beyond Ailsa Craig to see the northern tip of Ireland.

CAMELLIAS

Originally from China, Japan and south-east Asia the camellia family comprises over 250 varieties, including *Camellia Sinensis*, the tea plant. A noted winter-flowering plant, the camellia became popular during the Victorian period, with many stately homes adding it to their collections.

Culzean

ADVICE

There is an entry charge for Culzean. Paths are generally made and fairly level. Two parts of the walk involve stairs and there is a scramble over a rocky shore to see the caves. This section can only be walked at low tide.

PARKING

There is a large car park at the castle. You can buy a combined house-and-park ticket or a park-only ticket should you want to stay outside.

START

The walk starts at the Gas House, which is situated in a little bay below the castle.

CONTACT DETAILS

Culzean Castle and Country Park, Maybole, South Ayrshire KA19 8LE
t: **01655 884455**
w: **culzeanexperience.org**

Ordnance Survey Explorer Map number 326
© Crown Copyright 2008

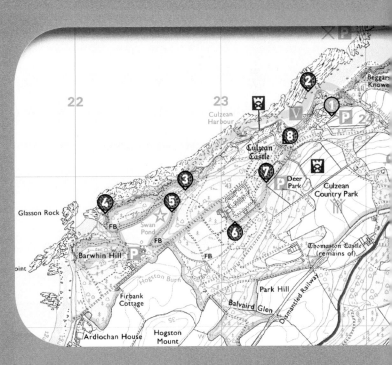

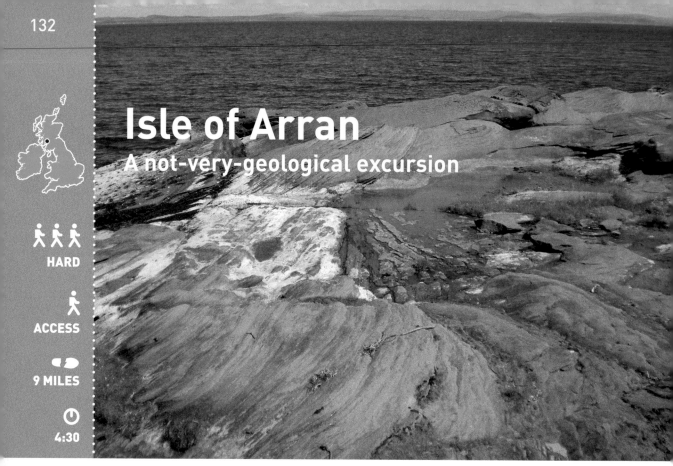

Isle of Arran
A not-very-geological excursion

HARD

ACCESS

9 MILES

4:30

Arran, in the Firth of Clyde, is the most southerly of the large Scottish islands. Most guides to the isle concentrate on its spectacular richness of geological features. This walk also covers other points of interest.

❶ The Isle of Arran is 19 miles long, 10 miles wide and 2867 feet high at Goat Fell. The highland boundary fault runs across the middle of the island so that the north is, geologically speaking, in the Highlands, while the south is part of the lowland Midland Valley.

Start the walk at **Lochranza Castle**. There has been a castle on the promontory here since the early 13th century, though what remains today is mostly from a 16th-century rebuilding. In the 1490s it was one of James IV's bases in his fights with the Lords of the Isles and Cromwell occupied it in the 1650s: it has been empty since the late 18th century and is now in the hands of Scottish Heritage.

❷ Walk along the main road towards the head of the loch, passing the Youth Hostel and the Church of Scotland. Turn left along a minor road with two Arran Path Network

signs: Fairy Dell/Ossian's Cave and Cock of Arran/Laggan. The road crosses a small stream and then bisects the golf course. At the T-junction at the end, turn left, signed Fairy Dell/Ossian's Cave – you'll be returning from the right-hand side at the end of the walk. A fence across the full width of the island just north of Brodick confines red deer (mainly) to the northern half: the high deer fences around all the houses in Lochranza can look a little incongruous until you see the cause, grazing by the roadside.

Follow the shore road until its end, then continue on a good gravel path along the **raised beach**. This was the shoreline until about 6000 years ago, but is now high and dry as Arran rebounds from the weight of glaciers in the last ice age. By Newton Point the view down Kilbrannan Sound opens out, and a view indicator points out places on Arran and the mainland.

❸ On the foreshore about half a mile after Newton Point is Hutton's Unconformity – you can't escape geology altogether. The path as far as Fairy Dell is easy and well signed. From here on the going will be harder and those wanting a short stroll can return to Lochranza over the hill: the path is signed for the Craft Centre at The Knowe. The coastal path continues with stepping-stones across the Allt Mor burn. Shortly after Fairy Dell is the rock fall at An Scriodan. The path continues, after a fashion, through the rocks, but if the tide is out it is much easier to walk along the shore instead. This is now the **Cock of Arran**, although there is no particular feature to mark it. Loch Fyne is now clearly visible to

the north, with Ardlamont Point to the right of its entrance. Ossian's Cave is well hidden just beyond An Scriodan among scrubby undergrowth, optimistically categorised by the Ordnance Survey as 'woodland'.

❹ Around here the shore changes from sandstone to limestone (more geology), which includes coal deposits. A **coal pit**, now filled with water, can be seen on the left of the path (the path itself is surfaced with broken coal for a few feet) and a little further on are the ruins of a salting house. The Hamilton family, who owned much of Arran, ran a salt works, Duchess Anne's Salt Pans, here from 1710 to 1735. The remains are still clearly visible, although rather overgrown with bracken in the summer months. Seawater was pumped in from the adjacent rocky inlet and then evaporated in a large iron pan over a coal fire. The salt was then shipped out from a small artificial harbour: a short pier and then two semi-circular quays enclosing a small lagoon. The remains and plan of the harbour can still be clearly seen at low tide. The buildings on the right as you head east were workers' cottages and stores. The final building on the left is believed to have been the salt excise officer's house.

❺ Shortly after the salt pans the path turns sharply away from the shore for a short distance. This is easy to miss, but it's just a short dead-end if you do. Otherwise the route is clearly marked. The rock here includes grits used to make millstones – one lies in the path – and forms for making the iron tyres on cartwheels.

HUTTON'S UNCONFORMITY

Sandstone beds from the Kinnesswood Group of the Lower Carboniferous period (about 360 million years old) lie directly on top of Dalriadian schists, which are about 520 million years old and have been heated and folded since being laid down (like the sandstone) as sediments. James Hutton, the 'father of modern geology', started to develop his 'Theory of the Earth' following observations made on the Isle of Arran. This led to the acceptance that the Earth was very much older than hitherto thought (to allow time for the lower schists to form before the sandstone).

THE CLEARANCES

Originally there were three farms on the north Arran shore: Laggan (above Laggan Cottage), Cuithe (near the path above the salt pans) and Cock (half a mile further to the north-east). These were worked on the traditional run-rig system and supported over a hundred people. All these people were evicted in 1829 to make way for two sheep farms, at the Cock and – about 10 miles to the east – Sannox. These too dwindled in turn and the last shepherd left Cock Farm in 1912, leaving the northern shore as it is now, uninhabited for 15 miles. Uninhabited by humans, that is: seals live on the shore and basking sharks, porpoises and the occasional dolphin can be seen just offshore.

At **Laggan Cottage** you have a choice of routes back to Lochranza. Along the coastal way, straight ahead, it's a mere 55 miles. For the rather shorter circuit turn right and head back via the marked Escape to Lochranza. This begins with a very steep section to about 300 feet, but after that becomes much more relaxed with some spectacular views.

6 On a fine day **Loch Fyne** can be seen to the north, with the village of Lochgoilhead in the distance. To the east (right) is Ardlamont Point and then the entrance to the West Kyle of Bute. Bute itself is the long island to the right. Like Arran it has striking variations in landscape, with a wild and uninhabited north and south and a low, fertile central belt. The uninhabited island of Inchmarnock lies just off the Bute Coast. Down the hillside are the remains of several farms, cleared in the early 19th century to make way for sheep.

7 The path you are on was originally built around 1850 by a Mr Cowie, partly for access to Cock farm and also for a slate quarry, which he opened near the summit. The slate produced was inferior and the quarry did not last long, but the spoil tips can still be seen. For many years in the 19th century the children of Cock farm had to walk this path to and from school in Lochranza every day – except in the depths

of winter, when a student teacher would stay at the farm and teach them there.

The path is now generally good, with occasional broken patches as you head downhill. Lochranza Distillery, opened in 1995, appears tantalisingly close up ahead and the spectacular **Torr Nead an Eoin** seems poised to break over Lochranza like a wave.

8 The rest of the route is straightforward so down the hill until the **path joins a track**. The main road from Brodick, the A841, can be seen descending from the left. It's a comparatively recent road, built in 1843. Before then Lochranza and the north-west of Arran were almost unreachable by land from the rest of the island.

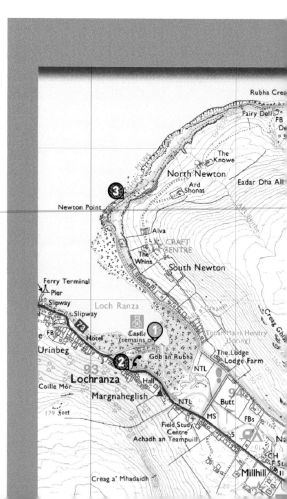

The track rejoins the minor road where you originally turned left, so another left turn takes you back through the golf course, over the stream and onto the main road. Turn right to return to the castle or left for the distillery, visitor centre and tearooms.

Isle of Arran

ADVICE

Ferries run from Claonaig on Kintyre to Lochranza and from Ardrossan to Brodick. They can be very full in the summer and advance booking is strongly recommended. Ardrossan has a frequent train service to Glasgow.

If you want to avoid a lot of scrambling past An Scriodan rocks, it is best to leave Lochranza at least one hour after, or three hours before, high tide. In the summer, midges can be a problem near sunset.

Lochranza has several B&Bs, a Youth Hostel and one hotel. The distillery visitor centre has a tearoom and there is a sandwich shop at the pierhead. The caravan site (more or less opposite the distillery) has a small shop, but major shopping is best done in Brodick.

PARKING

There is plenty of free parking at or near Lochranza Castle.

START

On the main road right by the castle.

CONTACT DETAILS

Isle of Arran Tourist Information Centre, The Pier, Brodick, Isle of Arran KA27 8AU
t: 01770 303776
f: 01770 302 395
e: info@ayrshire-arran.com
w: ayrshire-arran.com

Ordnance Survey Explorer Map number 361
© Crown Copyright 2008

Glasgow
The River Clyde walkway

MEDIUM

ACCESS

2 MILES

2:00

GLASGOW DOCKS

Four great docks were constructed during the second half of the 19th century: the Kingston Dock (where Kingston Bridge now stands), Queen's Dock (now home to the Scottish Exhibition and Conference Centre), Prince's Dock (filled in) and Rothesay Dock at Clydebank. The latter was built to handle exports of coal from local collieries and imports of iron ore for the steel industry at Motherwell.

From war memorials and suspension bridges to ladders and cranes, this urban walk takes in the eclectic nature of a city moulded by a river.

❶

You start your walk at the **Glasgow Science Centre** on the south side of the river. Now a shiny-silvery building emblematic of the 'new Glasgow', this was previously the location of the Queen's Dock.

The Queen's Dock was really at the hub of activity on the Clyde a century ago. As historian John Riddell has written, 'The 80 years leading up to the First World War saw Glasgow Harbour undergo massive expansion. In this time the riverside quays were extended along both the north and south banks of the Clyde as far downstream as the neighbouring burghs of Govan and Partick. By 1914 the quays following the line of the river provided nearly 11 miles of accommodation for all types of shipping.'

Here there were huge engine works and the sound of riveting would have been deafening. Ships would have been moored all around here and there would have been the noise of hooters and tugs turning.

The building on the left – Napiers Yard, famous for shipbuilding – would have housed the mechanism for the dry dock.

② Cross the river on the Millennium Bridge (or the Bells Bridge next to it) and turn left along the walkway, where you will come to the **SV** *Glenlee*.

By the 1890s the Clyde was the greatest shipbuilding river in the world. The *Glenlee* was a typical workhorse of the period with simple rigging, cheap to operate and fantastically dangerous. Built in 1896, the *Glenlee* circumnavigated the globe four times, carrying such noxious cargoes as guano for fertilizer. But the period when ships like this were at their prime was relatively short and by 1904 they were replaced by more powerful steam ships. Today the *Glenlee* is one of only five remaining Clyde-built sailing ships still afloat in the world.

③ Now turn around and retrace your steps along the riverside, past the Bells Bridge and on to the **Finnieston Crane**. This is a perfect example of maritime engineering. It was built for one purpose only – to load Glasgow-built railway carriages onto ships to be transported all over the world. Built in 1926, at the time it was the largest hammerhead crane anywhere in Europe; standing at 195 feet, and it was always in great demand.

Though it has been non-operational since the early 1990s it is still used for charity abseils and zip crossings of the river. During the 1988 Garden Festival a huge locomotive built of straw was suspended from the crane as a nostalgic reminder. It is one of the emblems of Glasgow, partly for its iconic status in the city's industrial past and partly because of the way it still dominates the skyline.

④ Carry on along the riverside until you come to the **Rotunda**. In 1895 this was the entrance to the only tunnel under the Clyde. It had three passageways – two for horse-drawn traffic and one for pedestrians. The horses and carts would be drawn onto lifts, which would take them down to the tunnel. They would then trot through and onto a lift at the other end. In each tower there were three hydraulic hoists for up traffic and three for down.

The Harbour Tunnel opened for business on 15 July 1895 during the Glasgow Fair holidays and traffic was light for the first week. On the following Monday, however, 218 vehicles used the tunnel. The next day it was 272 and the secretary of the company that had supplied the machinery reported, 'The horses generally have taken most kindly to the lifts, and are carried up and down without trouble. Carters said that by avoiding the steep inclines at the nearby ferries they could take five extra bags of flour per journey.'

As cars took over from the horse and cart the costs of running the Rotundas became more than the revenue and they were eventually closed. The pedestrian tunnel was closed on 4 April 1980. In 1986 both the vehicular tunnels were filled in.

⑤ Carry on walking along the riverside. As the settlement of Glasgow grew up along both banks of the river, communication between the two sides became essential. The most obvious way to cross the water was by ferry and there was certainly a regular service operating at Renfrew by the 17th century.

Another service sprang up between Govan and Partick, with the impetus coming from cattle drovers wishing to cross the river to reach their markets, although it seems hard to believe that two of the most industrialised areas of the country once had cattle passing through the streets.

THE ROTUNDAS

Though briefly re-developed during the 1988 Glasgow Garden Festival as the 'Dome of Discovery', the South Rotunda has since been derelict awaiting improvement. The North Rotunda was used as a casino and is now a Japanese Restaurant.

SCOTTISH BRIGADE

The surviving members of the International Brigades left Spain in early 1939 when a Fascist victory became inevitable. La Pasionaria said this at their farewell, 'Comrades of the International Brigades. Political reasons, reasons of state, the good of that same cause for which you offered your blood with limitless generosity, send some of you back to your countries and some to forced exile. You can go with pride... We will not forget you; and, when the olive tree of peace puts forth its leaves, entwined with the laurels of the Spanish Republic's victory, come back!'

THE GORBALS

On the southern side of Glasgow stand the tower blocks of one of the most famous districts of the city, the Gorbals. The traditional area for new immigrants to settle, the Gorbals has played host to many newcomers, including the Irish, Chinese, Jews and Lithuanians from the Russian Empire. Today it sees a new influx of the young and wealthy who flock to buy flats in an area undergoing a period of intense renovation.

However, it was with the rapid growth of the 19th century that the ferries really took off and by the end of the century there were eight in operation, mainly carrying shipyard workers from their homes to the yards and back again. In fact, some of the ferries went directly to jetties at particular shipyards.

Inevitably, ferry traffic declined along with the shipyards and in the 1960s the extension of cross-river road traffic via the Clyde Tunnel and **Kingston Bridge** sounded the death-knell for the river craft. The final ferry to cease operations was the Kelvinhaugh Ferry in 1982.

6 With the prosperity and opportunities brought by the flow of a great river come all the attendant dangers and Glasgow has had its own system for helping those in trouble on the river for over 200 years.

In 1790 the Glasgow Humane Society was formed to rescue those in difficulty on the Clyde. The need for an 'official rescuer' arose from a problem in Scots Law at the time. Suicide was a criminal act and many bystanders, seeing someone in trouble in the water, would be loathe to intervene in case they were deemed to be an accessory to the crime.

A full-time rescuer was first appointed in 1859. It is estimated that Ben Parsonage, who held the post between 1931 and 1979, saved over a thousand lives during his tenure. The society today is largely funded by Strathclyde Police and by public donations, and still operates from a house near the suspension

bridge and by the sign that says **Rescue Ladder N170**. It is the oldest organisation of its kind in the world.

7 As you approach Custom House Quay, you will see **La Pasionaria Memorial**, one of the few memorials in the country to those who left Britain to join the International Brigades to fight in the Spanish Civil War in the late 1930s. Of the 2000 or so Britons who fought in Spain, around a quarter were Scots. Of the 134 Scottish Brigaders who lost their lives in Spain, 65 were from the city of Glasgow.

The memorial was sculpted by the Communist artist Arthur Dooley in the mid 1970s and was unveiled in 1977. The figure represented in the statue is Dolores Ibarruri, better known as La Pasionaria, a fiery female politician from northern Spain, who was a popular figure within the Republic and well-known abroad for her rousing speeches in support of the Republic. A quotation from La Pasionaria is written on the base of the statue: 'Better to die on your feet than live forever on your knees.'

8 The final part of the walk takes in the **South Portland Street Suspension Bridge** and allows a great opportunity to see the city as a whole. It has often been said that 'You can see all of Glasgow here', and, while not literally true, you can certainly get a strong sense of the city.

The story of the city begins by looking at the Fishmarket Tower, which forms the boundary of the medieval city, before the dramatic increase in size experienced during the mercantile period and the Industrial Revolution.

On either side of the river stand two buildings that attest to the impact of

immigration on the city. On the north bank stands St Andrew's Cathedral, the focal point for Scotland's Catholics, most of whom are descended from Irish immigrants who arrived on the Broomielaw Quay throughout the 19th century. On the other side of the river stands the Central Mosque, the centre for Scotland's Muslims, many of whom have settled in the city since the end of the Second World War.

To get back to the starting point at the Science Centre, simply retrace your steps and cross the river via the Bells Bridge or the Millennium Bridge a little further along.

Glasgow

ADVICE
This is a flat walk along pathways and is mostly suitable for pushchairs and wheelchairs.

Ordnance Survey Explorer Map number 342
© Crown Copyright 2008

PARKING
There is ample pay-and-display parking at the Science Centre.

START
The walk starts at the Glasgow Science Centre on Pacific Quay on the south side of the River Clyde.

CONTACT DETAILS
**Glasgow Tourist Information Centre,
11 George Square, Glasgow G2 1DY**
t: **0141 204 4400**
f: **0141 221 3524**
e: **enquiries@seeglasgow.com**
w: **seeglasgow.com**

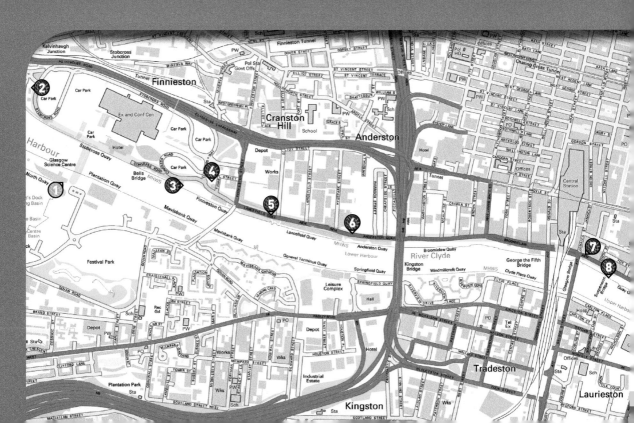

Mull
Way out west

MEDIUM

ACCESS

8 MILES

3:30

To most visitors, the Ross of Mull and Fionnphort are simply places to pass through quickly on the way to or from Iona. This walk offers a chance to explore some of the lesser-known history of this beautiful area.

❶

Start at the ferry terminal in **Fionnphort** – this is also the bus stop. Walk back through the village and take the turning on the right signed for Fidden and Knockvologan. You follow this minor road to the end, a distance of about 3 miles. Shortly after you pass through a gateway there is a large outcrop of pink granite on your right. This is the first clear example of the stone for which quarries all over the Ross of Mull were opened – one of these quarries, on Erraid, is at the far end of this walk. As you get to Fidden farm, Erraid is clearly visible about

a mile ahead. The quarry is now overgrown, but the rows of cottages built for the workers there can be easily seen, as can the small, light-green observatory on the hillside above and to the right of them.

About a mile after Fidden farm the road turns sharply to the right, with a track continuing straight ahead on the other side of a gate. At the gate are the ruins of a granite-built blackhouse. After another mile you come to Knockvologan (Milligan's Hill) farm. There is car parking here, so if you have a car available you can miss out the on-road part of the walk.

2 Knockvologan is the start of a series of walks on the Tieragan Estate. This is now owned by a charity, Highland Renewal, which is encouraging the re-establishment of native woodlands and wildlife. Leaflets giving details of walks are available at the entry gate.

The walk continues down the road, through the gate with blue floats on the posts. At the next farm make a short detour by turning right and following the grassy track for about 200 yards along the hillside. To the south you have an excellent view of the Torran Rocks, an area with a well-deserved and dreadful reputation for shipwrecks.

Back at the farm, take the track down the hillside – it has a concrete surface for the first 10 feet or so. At the first beach you can see two **posts sticking out of the sand**, each with a solid iron hook beside it. These are the visible remains of a ship's lifeboat. The wooden posts are the stem (front) and stern (back), and the hooks were used to lower the lifeboat into the sea. The intention was that when a lifeboat entered the sea and was lifted by a wave, both hooks would lift free of the ropes by which they had been lowered. Unfortunately this did not always work, and occasionally a lifeboat would be left hanging by one end and all on board dropped into the sea.

3 Follow the path across the beach and through a natural cutting in the rocks. There is a very impressive pink granite wall to your right. Pass a modern boathouse as you emerge onto a smaller beach. The inlet to the south of you is known as **Tinker's Hole** and is one of two excellent sheltered anchorages

for yachts in the Sound of Iona (the other is Bull Hole, just north of Fionnphort).

Keep on the track, although it is now a little harder to distinguish, across the small headland, past or across a tumulus, and you arrive at the Sound of Erraid. This is crossable by foot for all but an hour or so either side of high tide. If it's still covered, now is probably a good time for a rest and a snack.

4 Cross the **Sound of Erraid** diagonally, aiming for the low area at the north end, just south of the hummocky promontory of Dun Aidean. At this point you have a choice. The easiest way to go is round the shore, hugging it closely if the tide dictates. Round the corner from Dun Aidean you pass a small jetty and a house on the shore. Cross to the far side of the bay, to the house, where you will see the track head up from the shore.

If you prefer, or if the tide is still high, you can cross the island instead. Follow the shallow gully up from the beach south of Dun Aidean and aim to keep the rocky outcrop ahead close to your right. Turn right and head down to the house, turning left in front of it to go round the bay and so on to the track. This route is rather boggy in places, and only recommended if the shore way is impassable.

5 The track leads on to the **remains of the granite quarries** – or, more accurately, to the lighthouse works, as these workings were established by Thomas and David Stevenson for the construction of Dubh Artach lighthouse. Pass in front of the cottage gardens. The cottages to your left were built for the skilled workers: quarrymen had wooden

barracks, now long gone. After passing in front of the cottages you reach a gate. Turn left up the track and, when you reach the gate just before the cottages, turn right up the steps to visit the quarry. The pointed metal structure lying here is the original air vent from the roof of the observatory, replaced when the structure was restored.

As you reach the level of the quarry, you will see a flat area to the right of the track. This is believed to be where trial assemblies for the lighthouse were carried out. As each course of interlocking granite stones was finished it was assembled here. When the fit had been checked and any final adjustments made, it was dismantled again and taken to the pier to be shipped out to the construction site. Conditions there could be dreadful, so it was important to make the job of assembly as straightforward as possible.

6 The quarry itself is now heavily overgrown, but the spoil heap is clear – you walk across it – and elaborately decorated with a stone labyrinth by the current island community. Beyond the spoil heap is a lone ruined cottage. The path passes to the left of this and then bears left up the hill. Below a rocky outcrop on the right it makes a hairpin turn up the hill, with granite steps up the first, steep, section. After a short but steep climb you reach the **lighthouse observatory**, which has been recently restored.

This round, iron-clad building has two windows, pointing at Dubh Artach and Skerryvore lighthouses. You may be able to see them on the horizon with the naked eye,

although binoculars help a lot. Every day the lighthouse keepers would hoist a ball on a mast at the top of the tower to show that all was well. If this was not seen from the observatory, the shore crew would hoist a signal. You can see the foundations of the mast beside the observatory. If no satisfactory reply was received a boat would be sent out to investigate.

There is an excellent view from here of Iona and the islands beyond. The dark lump visible over the north end of Iona is Staffa, well known for Fingal's Cave. Further away and to the left of Staffa are the Treshnish islands. To the north-east is the spectacular Ardmeanach peninsula on Mull, with striking exposed horizontal strata.

7 Return the way you came, through the quarry, past the end of the cottages and go on to the **harbour**. On your left is the original water supply pump. The small stone hut immediately behind it is a covered reservoir. The harbour was, like everything else on Erraid, built for the construction of Dubh Artach lighthouse. This explains the solidity of the masonry work – the sea wall on the west side is particularly impressive. At the far end of the quay the remains of a crane can be seen. This was used to lower the finished blocks of stone into boats for transport to the lighthouse itself.

Please note that, like the cottages, the harbour is now used by a part of the Findhorn Community. Please respect their privacy.

8 You have a choice of three ways back to the start. If the tide is low, you can be adventurous and cross the sands from near the harbour at Erraid back to Fidden farm. You can take the straightforward route back by retracing your steps all the way to Fionnphort. Or if you are enjoying yourselves you can take the longer route back by retracing your steps as far as the ruined blackhouse and then turning right through the gate. This track leads you round the bottom of **Loch Poit-na-h'I**, through Pottie farm and onto the main road about 2 miles outside Fionnphort.

Mull

ADVICE

At certain times of the year Erraid may be inaccessible around high tide. As conditions vary it is best to take local advice, but as a general guideline you should have no problem crossing more than two hours before or after high tide. Fionnphort has several guesthouses, a shop, toilets and a pub/restaurant. There are further facilities a short ferry ride away on Iona.

PARKING

Fionnphort is about an hour by car from Craignure ferry terminal. There is a pay-and-display car park at the ferry terminal in Fionnphort. The town is also served by buses from Tobermory and Craignure.

START

Start at the ferry terminal in Fionnphort.

CONTACT DETAILS

Tobermory Tourist Information Centre, Main Street, Isle of Mull, Tobermory, Strathclyde PA75 6NU
t: 08707 200 625
f: 01688 302145
e: info@tobermory.visitscotland.com
w: visitscottishheartlands.org

Isle of Skye
Not for the fainthearted

HARD

ACCESS

5 MILES

4:30

LANDSCAPE PHOTOGRAPHY

Good landscape pictures often have great depth, with items of interest in sharp focus from the foreground through to the far distance. Traditionally this is achieved by tilting the lens or camera back and by stopping the lens down to some extent. If you see someone at Ord with a large wooden camera trying to compose an image upside-down on a glass screen, this is probably what they're trying to do.

This wild walk takes you off the beaten track, providing you with the chance to test your navigational skills, spot some wildlife and have a picnic on a quiet beach with fantastic views.

Park overlooking **the beach at Ord**. This area is very popular with artists and there are several signs for artist's studios on the minor road down to the sea. Just beside the car park there is also a garden, Anacarsaid, which is open during the summer months. The view from this car park is ever changing as the clouds come in from the west or build over the distant Cuillin hills. The area is also a favourite with landscape photographers as it has an interesting stream, seaweed and rocks on the beach, tranquil sea and dramatic ranges of mountains to lead right into the scene.

2 From the car park head south over a small bridge, crossing the stream that reaches the sea at this point. Turn left immediately over the bridge and go through a series of three fields along a **track** tending to bear right uphill. When the path reaches a wire fence and turns sharp right, go straight on, crossing the fence and heading through boulder-strewn woodland with ash and hazel trees and up onto the open hillside. If you have only been accustomed to walking on well-trodden paths this may come as a bit of a shock, as you are now boldly going where few have ever gone before. The route is quite easy, especially on a summer's day with good visibility, but it is essential to carry a detailed map and compass, and know how to use them, as the weather can close in at any time of year and you may need to find your way back in cold, foggy conditions even if it starts out sunny. Walking across a heathery hillside can be more pleasant than the constant bone jarring that comes with well-worn rocky tracks, although watch out for holes in the hummocks as it's quite easy to twist an ankle.

3 The route now heads south-west across the open hills with marvellous views of **Tokavaig woodland**, jagged black Cuillin and out to Rhum and other islands. Keep generally to the broad ridge for the first half-mile of this section. Then take a small detour south-east to avoid a steep gorge that crosses the route, soon returning to a south-westerly course into the boggy col behind Sgiath-bheinn Tokavaig. Keep to the edge of the low saddle to avoid the wettest areas.

One of the plants found on these hillsides is common butterwort. It has a rosette of yellow-green sticky leaves about an inch wide and 2½ inches long, and quite an unusual violet flower an inch across on a long stalk. There are not many nutrients in the soils

so the plants capture insects on their sticky leaves and dissolve them to provide extra food. Besides common butterwort, pale butterwort also grows in this area. It has smaller pale lilac flowers, and its distribution is much more restricted than common butterwort growing in north-west Scotland and parts of south-west England. Another insectivorous plant that grows in the more boggy areas is sundews. In this case the leaves have sticky globules on stalks to trap and dissolve the insect.

4 From this high point the route is north-west, downhill, generally following a stream's course. It should be quite easy to spot the **stream**, but keep well to the right as it plunges into a steep ravine carved through soft rocks. The walk as a whole circles round a very rich area for plant life in the Tokavaig woodland and ravines, with many unusual species of mosses, ferns, lichens, orchids and other plants. Much of the interest centres on

⑤

the combination of warm damp conditions supplied by the Gulf Stream, sheltered ravines and stunted forest, together with the limestone rock and alkaline soils, which are rather different to the more acid conditions found in many other parts of the north-west. The ravines can be rather dangerous so it's best to resist the temptation to lean over and instead look at the luxuriant covering of lichen on the hazel trees beside the road further on.

⑤ On the way down the hillside it is necessary to cross a deer fence a couple of times. The first crossing is easy as there is a gate, but at the lower crossing you may need to climb over. Find a strong place close to a wooden pole to help you cross and avoid damaging the fence. It is actually not difficult to cross, even though the high fence may look daunting at first. Keep going down the hillside then shortly before reaching the road there is a grassy area with several species of **orchids**. Continue down through a narrow band of woodland and turn left at the road. On the roadside here are several types of fern, including lemon-scented mountain fern, which has a distinct lemon smell if you crush the fronds. Carry on along the road for a couple of hundred yards.

⑥ When the road turns left towards Tokavaig, carry straight on along a track heading towards the ruins of **Dunscaith Castle**. There is not much of the castle left, but it's still possible to cross a stone bridge and climb up steps onto the rocky promontory and see the crumbling walls. The castle had a turbulent history among the clans in the 14th and 15th centuries, finally being captured by James I of Scotland. The MacDonalds were actually allowed to retain the castle, although it was finally abandoned in the 17th century and fell into ruin. The castle is also a good location to stop, have lunch and watch out for otters, seals and a range of sea birds.

⑦ Go down from the castle onto the **rocky beach** and head north-east back towards Ord. The cliffs here have some rather unusual plant communities and are well worth a look. It may seem rather strange, but plants at the base of the cliffs at the back of the beach here include bluebells, a woodland plant, growing alongside thrift, a seaside plant, and skullcap a freshwater plant, not to mention English stonecrop (growing in Scotland) and the dense 'beard' of shrubby lichen covering the rocks.

⑥

⑦

8 Walk along the beach for about half a mile until reaching a **rock arch**. Just after this point it is probably best to scramble back up the cliff as it's possible to get cut off by the tide when continuing on the beach route beyond here. The scramble is not difficult, but if you are unsure about it then retrace your route through the arch as there are several easier ways up the cliff on the grassy slopes.

Once on top of the cliffs walk through the heather all the way back to Ord. Just before getting back you may need to go through a small area of woodland and across a sheep field.

Isle of Skye

ADVICE

The route is classed as hard, because it is a 'new' route without a path, so you will need to be experienced at navigation and have a map and compass. You will also need to be reasonably fit and have good boots, as the route is a little boggy in places and is through hummocky heather, which can cause sprained ankles etc if you are not used to this kind of walking. On a clear day this should be quite easy and exhilarating for experienced walkers. For those with restricted mobility there is an excellent beach and garden walk at the start point of the main walk and at Tokavaig halfway along the main walk (you can drive to Tokavaig). Also from Tokavaig there is a half-mile walk out along a track to a ruined coastal castle. The track is rather uneven and boggy in places, but it may be suitable for large-wheeled baby buggies.

There are no toilet facilities or refreshments.

Ordnance Survey Explorer Map number 412
© Crown Copyright 2008

PARKING

There is a car park overlooking the beach at Ord.

START

Start in the car park by the beach at Ord.

CONTACT DETAILS

Broadford Tourist Information Centre, The Car Park, Broadford, Isle of Skye, Highlands and Islands IV49 9AB
t: **01845 22 55 121**
e: **info@visitscotland.com**
w: **visithighlands.com**

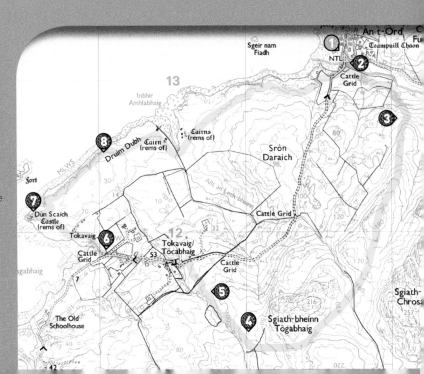

Rubha Reidh
A walk on the wild side

HARD

ACCESS

4 MILES

3:00

SHINING LIGHT

Rubha Reidh lighthouse was built by David A. Stephenson and opened in 1912. The light is 120 feet above sea level and has a brightness of half a million candlepower. It was switched from lighthouse keeper-operation to fully automatic in 1986.

If you like your nature wild and unspoiled, dramatic red cliffs, small sandy beaches and unusual flora and fauna all feature on this walk, which starts in a remote spot 45 minutes' drive from the nearest village.

This is a wonderful walk through a beautiful landscape, but before you embark on it please note that the cliffs and paths are crumbling all the time, so in some places it's best to avoid the path that's right on the edge. The wind is often strong and gusty, too, which is another good reason for staying away from the edge.

1 From the lighthouse head east towards the obvious **gap in the rocks** at the last hairpin bend on the road. The broad track heads out through huge red slabs of bare sandstone rock, which go down into the

roaring waves below. Soon the reason for the track's existence becomes obvious, as you can see the remains of a concrete jetty among the jagged cliffs, although it would seem almost suicidal to attempt to launch a boat here.

2 The gales that frequently batter this part of Britain have had a severe effect on the vegetation, which in many places is only a few inches tall. However, if you look into he sink holes or any sheltered ledges, there is a luxuriant growth of ferns and other plants, including a very local species called Scots lovage, which only grows on cliffs around Scotland and in a few other places in north-west Europe. With its umbels of white flowers and leathery green leaves it looks a little like a small angelica plant, but it only grows 12 to 24 inches tall.

Another unusual species to look out for is roseroot, a member of the stonecrop family, which grows on these sea cliffs, but also in high mountains. It has thick grey leaves, often with a purple tinge. Its flowers are yellow and it bears bright red fruits, which are often mistaken for flowers. It grows in a clump and reaches about 12 inches.

3 Head back up the track a short distance to cross the stream and continue east along the coast. This is one of the places where you should avoid the paths on the very edge of the cliff, which can readily disintegrate. Also watch out for partly covered **sink holes** in the hillside, where water has eroded relatively weak sections in the underlying sandstone rock. These are the beginnings of deep chasms like the one by the jetty.

There is a steady stream of seabirds along this section of coast, often flying at about eye level. Look out for greater black backed gulls, great skuas and gannets. If you're lucky, the gannets may put on a show by plunge-diving for food just off the coast. Note that besides the adult gannets with their predominantly white plumage and black tips to the wings, there may also be juveniles, which have a much darker appearance. All these birds are surprisingly

large, with wingspans ranging from 51 inches for the great skua to 70 inches for the gannets. By comparison, the black headed gulls most city dwellers are familiar with from their local parks have a wingspan of just 39 inches.

4 Follow the coastline on one of the many tracks, heading inland and uphill to get round any difficult sections, such as where the stream and steep valley cross your route at about three-quarters of a mile from the starting point. Looking out across the hillsides you may initially think they are heather moors, which in other parts of Britain are dominated by just one species of heather (*Calluna vulgaris*), but if you look more closely there are a whole range of species. One of the most obvious is crowberry, with its linear, glossy green leaves about a quarter of an-inch long and black berries. In this species there are male and female plants, but only the female ones have berries. There are also the small blue flowers of milkwort, the yellow tormentil, the strange hooded pink flowers of lousewort, and the spotted pinkish-white flowers of heath spotted orchid. Other species on the hillside include patches of wiry white cladonia lichen. This only grows an inch tall, so other plants can easily smother it in other habitats, but in these severe conditions

GREAT SKUA

These thickset brown seabirds can initially be confused with immature gulls, but on closer inspection you can see their heavy, meaty features and they have a white flash towards the end of their wings. They are pirates, who will chase down other seabirds and force them to disgorge their food, but they also scavenge or seize fish that are close to the surface. In Britain they breed on Orkney and Shetland, where they dive to attack any potential predator, including humans, often delivering a blow with their feet.

ROCK OF AGES

The rocks you cross when you come to Gairloch and on to Rubha Reidh are some of the oldest in Europe. Firstly there are the Lewisian gneisses around Gairloch, which are in the region of 2500 million years old – more than half the age of the planet. However, most of the peninsula out towards Rubha Reidh is Torridon sandstone, which is somewhere in the region of 1000 million years old and probably too old to contain any fossils. There is also a small area of new red sandstone.

⑤

where all plants are under stress it finds plenty of bare areas to colonise. If you are there on a dry sunny day then the lichen will feel hard and dry, but under damp conditions it quickly softens up like a sponge. If you look closely at the lichen it appears to be a miniature tree, an observation that is not lost on model-makers and toy train enthusiasts.

⑤ After the stream you should start to get good views of the beach and rock stacks of **Camas Mor** far below. Continue in an easterly direction for just over half a mile, avoiding the cliffs by staying inland of them. Then descend towards the coast at the end of Camas Mor beach.

⑥ At this point you can enjoy the beach and then return by retracing your route. While you are walking on the beach look out for seaweeds. One of the easiest to identify is channel wrack, a small brown species with narrow branching fronds up to 6 inches long, their inrolled margins forming a kind of channel. It's often the species that grows highest up on the rocks in what is a very

hostile environment for seaweeds. However, it has a number of adaptations to survive the conditions, such as the rolled fronds and a fatty layer over the cells, both of which reduce water loss by evaporation. It is also able to survive on less nutrients than other seaweeds, but this does mean that it grows slowly and remains relatively small.

⑦ If the tide is low, it is possible to go along the beach, through a natural arch, along another beach and then scramble up the cliffs at the stream you crossed earlier in the walk. Otherwise, enjoy the first part of the beach then retrace your steps and return the way you came.

In the wetter areas one of the most obvious plants is **sphagnum moss**. There are several different species, some growing along the boggy streams and others in damp depressions. The moss has a network of hollow cells that draw up water to the growing tip, so the whole mass of moss is very wet. Indeed you can wring water out of it. Maintaining these very wet conditions means that the soil tends to be starved of oxygen, which inhibits the breakdown of organic matter. Thus plant remains are not fully degraded and gradually accumulate into layers of peat.

High levels of rainfall and cool conditions in this part of the world also contribute to the very slow breakdown of organic matter

⑦

and facilitate peat formation. With climate change it is not clear whether this area will continue to be cold and wet. If instead it warms up and/or becomes drier, then the peat may start oxidising and disappearing, in the process releasing large amounts of CO_2 and further accelerating the greenhouse effect.

The current severe climate, peaty low nutrient soils and sheep grazing also prevents the growth of trees.

⑧ To return to the lighthouse either retrace your steps along the coast or, if the weather is clear, you could try going over the open moors on a higher level route.

At the end of the walk it is worth spending some time at the area round the lighthouse itself, as it is very nice, with access to tiny **beaches of red pebbles** and all the interesting plant species visible on the low cliffs.

SANDSTONE FORMATION

Sandstone consists of sand and small pebbles held together by minerals, and the strong red colour of the sandstone here comes from fragments of oxidised iron. When this sandstone was being laid down, present day Scotland was part of a large continent south of the equator and far from the sea, so it is thought that the sand and pebbles were deposited by rivers that came from hundreds of miles to the west.

Rubha Reidh

ADVICE

Much of this route is on narrow cliff-top paths that are not suitable for those with limited mobility, although the first 200 yards are on a broad track. It is important to check tide times before attempting this walk, as one section is only passable at low tide.

START

This walk commences at Rubha Reidh lighthouse, which is now an outdoor centre and cafe, although it's only open on certain days.

PARKING

The lighthouse car park is rather small, but there is another small car park a few hundred yards before you get to the lighthouse.

CONTACT DETAILS

Rubha Reidh Lighthouse, Melvaig, Gairloch, Ross-shire IV21 2EA
t: 01445 771263
e: ruareidh@tiscali.co.uk
w: ruareidh.co.uk

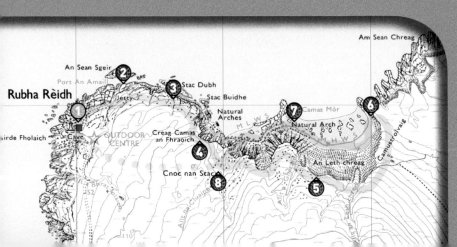

Ordnance Survey Explorer Map number 434
© Crown Copyright 2008

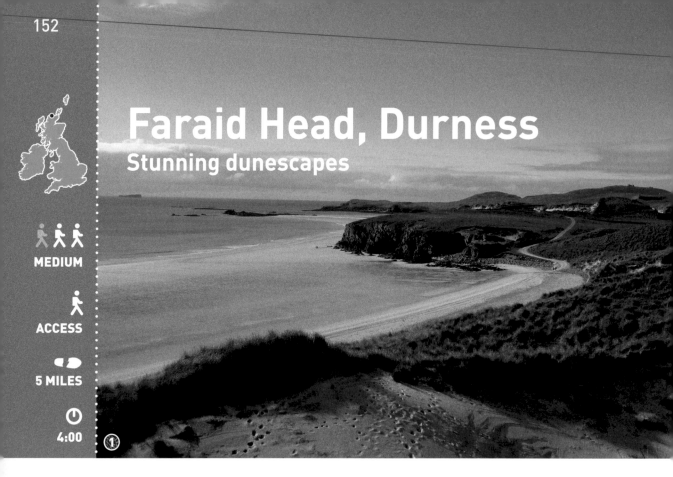

Faraid Head, Durness
Stunning dunescapes

MEDIUM

ACCESS

5 MILES

4:00 ①

CONSERVATION

Some of the principal reasons for Durness' designation as a European Special Area of Conservation (SAC) are its sand dunes, limestone pavement, calcareous arctic alpine flora and calcareous lochs. The SAC covers all of Faraid Head and a large area to the west and south of Durness.

The coastline around Durness has been designated a special area of conservation and you'll encounter a wide range of unusual plant life as you explore its striking system of sand dunes.

There are dune systems on several parts of the British coastline, but the Faraid Head dunes are some of the tallest and most mobile, with large, bare expanses and towering, unstable sand cliffs. This dune system may be so extreme because the dunes are not allowed to find their natural stable shape due to the road that leads through the middle and the constant battle of using heavy machinery to clear it of sand. The dunes are sited on an area of tall jagged rocks that form a peninsula, so their natural shape may be rather extreme anyway.

① From the car park head due north along the **beach** for just under half a mile. Unlike the dunes themselves, this is a nice firm, sandy beach, famed for its sunsets and suitable for swimming, although do note that the tide comes in rather fast across the flat sand. At the far end of the beach there are interesting dark, jagged, 'soapy' schist rocks that jut up through the light sand. Schists have a slightly sparkly appearance and when you touch the rock it's a bit like feeling a bar of soap, even though it's hard and dry. Compare this rock to the Durness limestone

that you'll come across later in the walk. It can be a similar colour, but has quite a different texture.

2 At the far end of the beach pick up a metalled track and head up into the dramatic tall **dune landscape**. This is an excellent place to study how plants can hold sand together. Look out for the tough, rope-like, far-spreading roots of the marram grass, but also its very fine network of roots that lie ust below the surface, stabilising the grains. These features are revealed where the dunes have eroded.

3 At the end of the dunes there is the choice of either following the track directly up to the military base or bearing left to walk round the cliff-tops, before joining up with the main track at the base. The latter option

adds about half a mile to the distance, but gives **good views of the cliffs**.

The grassland beyond the dunes is a special type called machair and is home to a species-rich plant community more usually seen on the Outer Hebrides, but also found in a handful of places on the mainland. The lime-rich, sandy soil is ideal for many plants, including several types of orchids, field gentian in both its normal purple colour and a white form, grass of Parnassus, white clover, buttercup and wild thyme. Looking down, there is a wide variety of different leaf shapes and flowers, but all are very small. As a comparison, the frog orchids here are about 1½ inches tall, compared to their more usual height of 4 to 6 inches.

4 At the military base turn right or east and follow the path along the fence, then along the cliff-tops up to a viewpoint marked by a cairn or pile of stones. The base is part of the Cape Wrath gunnery range. From the viewpoint it is clear how narrow Faraid Head is and you get the impression that one big Atlantic storm could wash all that sand away and turn the headland into an island.

5 There is a puffin colony at the base of the cliffs, about 270 yards south of the cairn. It's possible to see the puffins speeding in to their burrows with mouthfuls of fish, but resist the temptation to get too close to the cliff-edge as it may crumble away. There are several other types of cliff-dwelling seabirds,

BALNAKEIL VILLAGE

If you park at the car park in Balnakeil, a mile north-west of Durness, it's worth taking a stroll round the village itself, which features a number of craft shops, as well as the ruined church and Balnakeil House.

such as fulmars, around **Faraid Head** and the rocks offshore, although the colonies are not particularly large and you certainly need binoculars or a telescope to get good views.

There are also large colonies of northern marsh orchids in this area. These orchids have wonderful rich, reddish-purple flowers and look quite different to the pale pinky-

purple or white flowers of the heath spotted orchids with which they occur in the machir grassland. It is unusual to see the acid-loving heath spotted orchids alongside the marsh orchids, which generally prefer neutral to alkaline soils. Soils in this part of north-west Scotland are generally acid peat, but here the calcareous sand has blown on top of the peat and neutralised it.

6 Continue on along the cliffs and pick up a path that heads into the dunes. The first **dunes** encountered on the way back are very dramatic as they merge straight into the cliffs and give the impression that at any moment the narrow sand path could fall away into the crashing waves far below. In fact, the waves are often a milky white colour from the sand that is constantly being eroded. Turn right for the short distance on to the metalled track and head back towards the beach and car park.

7 On the return walk, look out for a large building to the left of the car park. Balnakeil House was built in 1744 and at one time was a seat of Clan Mackay. The **ruined church** opposite dates from 1619. There are a number of interesting graves in the church and its surrounds, and it is well worth a look. The small stream beside the car park is choked with monkeyflower, which is not native to Britain, but which now makes a brightly coloured nuisance of itself in many places.

8 The dunes and machir grassland along the walk contain a wide variety of interesting plants. However, for botanists there is another, even more special additional section to walk, which can be done from the Bainakeil car park, after the main walk. Go west onto the golf course keeping close to the seaward side and following a line of small white stones in the grass. Some of the **orchid** species from the Faraid Head walk are present on the golf course, but the plants are substantially larger. Resist the temptation to linger on the course, but instead use the gate to cross a fence marking the far end of the course.

The scenery changes from manicured greens to a very nice section of sugary

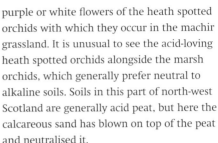

limestone grassland with many bare patches. Sugar limestone is so called because the rock breaks up into sugar-like granules. The habitat is very fragile and in most other parts of Britain the area would be out of bounds. However, here in the far north-west of Scotland there are few visitors and people are permitted to walk, but keep to the path where possible and don't abuse this privilege.

As well as the orchids from the Faraid Head walk, you'll see many other rarities, such as yellow saxifrage, but the most special plant here is the Scots primrose or, to give it its Latin name, *Primula scotica*. The Scots primrose only grows at a few sites along the north coast of Scotland and, not surprisingly, is the emblem of the Scottish Wildlife Trust. It's a small plant, with a flower spike of about 3 inches and a rosette of mealy white leaves. The flowers are purple with a yellow centre. The Scots primrose flowers from May to early June and from late July to August.

The trail continues for several more miles and features many other interesting plants, but the Scots primrose is a good point at which to turn back. As you return across the golf course don't miss the spectacular hole over the sea where most of the fairway is missing and instead there is only the tee, the green and a maelstrom of crashing waves in between.

Faraid Head, Durness

ADVICE
The walk is not recommended for wheelchair or buggy users. The route is open all year round, even when the naval gunnery range at Cape Wrath just less than 10 miles away is in use, but keep an eye out for military vehicles as it is sometimes difficult to hear them coming on the sandy road, and take care near the rock and sand cliffs, which may collapse without warning. The nearest toilets and refreshments are in Durness, a mile away.

PARKING
Park in Bainakeil car park about about a mile north-west of Durness.

START
The starting point is Bainakeil, where you park.

CONTACT DETAILS
Durness Tourist Information Centre, Durine, Durness, Sutherland IV27 4PN
t: 08452 255121
f: 01506 832222
e: info@visitscotland.com
w: visithighlands.com

Ordnance Survey Explorer Map number 446
© Crown Copyright 2008

Stromness
Gateway to Orkney

MEDIUM

ACCESS

4 MILES

3:00

GEORGE MACKAY BROWN

It was in a house at Mayburn Court that this famous Orcadian poet was to compose his greatest works, including *Fishermen with Ploughs* in 1971, the novel *Greenvoe* the following year and *A Celebration for Magnus* in 1987, with a musical score by the celebrated composer Sir Peter Maxwell Davies.

The beautiful, windswept islands of Orkney provide one of the most dramatic settings for a coastal walk in the British Isles and here, in Stromness, you will find the perfect opportunity to explore them.

❶

The walk starts in front of a merchant's house known as **Graham House**. Stromness and Kirkwall have always vied with each other as the two major towns on the 'mainland' of Orkney, but this rivalry has not always been friendly. In the 18th century, Scots Law decreed that only those towns designated as Royal Burghs could take part in foreign trade. Despite the fact that Kirkwall held the status of Royal Burgh, the majority of trade passed through Stromness.

The trading success of Stromness was not appreciated at the other end of the

island because Kirkwall had to stump up the tax liable from the Stromness merchants. Unsurprisingly, the Kirkwall residents decided that measures should be taken to redress this and in 1719 they came up with a plan.

Stromness merchants were ordered to make payments directly to Kirkwall, a system that was as unpopular in Stromness as the previous one had been in Kirkwall. This agreement held for over 20 years, until in 1742 one Stromness merchant, John Johnson, decided enough was enough and refused to make the payment, setting off a chain of non-compliance. Though Johnston died soon after, local merchant Alexander Graham stepped in and carried on his action.

The row carried on until it reached the House of Lords, which ruled in favour of Stromness 15 years after the protest began. But victory came at a price, Alexander Graham, who had led the campaign for over a decade, was now completely ruined.

2 A little further along Dundas Street you will reach Mayburn Court. The house facing onto the main road with the plaque on it belonged to **George Mackay Brown**, the man who dominated the story of 20th-century Stromness. Poet, novelist and playwright, Mackay Brown did more to promote the cultural life of Orkney than any other individual.

Born in Stromness in 1921, Mackay Brown enjoyed a passionate love affair with his home town. In 1968 he moved into Mayburn Court, a small council-owned property, which the great poet rented at the princely sum of 18s 6d (93p) a week. At the age of 47 this was the first time that he had lived alone, but despite his fear of moving,

the house was to become in many ways his muse – his 'watchtower' as he described it.

3 A little further on the road on the right is **Login's Well**. This watering hole links Orkney, and Stromness in particular, with the opening up of the New World. The story began in London in 1670 when Charles II's cousin, Rupert, and a band of nobles set out to form a 'Company of Adventurers' which, in time, was to become the effective ruler of large swathes of British North America.

This company was granted an exclusive Royal Charter to trade at the mouth of the Hudson, and so the Hudson's Bay Company was born. The company's ships needed a safe place to anchor and replenish stores, and it is on the wealth of the Hudson's Bay ships that Stromness grew. The last port of call before braving the Atlantic, this sleepy town became the place where the ships took on water before the open ocean, right here at Login's Well. The well continued to supply ships with water until it was finally sealed in 1931.

4 Heading south out of the centre of town, overlooking the entrance to the sheltered harbour, sits a solitary cannon pointing defiantly into the bay. Known as the **Liberty Cannon**, this is a reminder of the trading conflict between Britain and the United States between 1812 and 1814, when the British tried to keep their supply line of grain from Canada open and the Americans attempted to stop them.

An important tool in the American campaign was the privateer, an armed vessel

THE HUDSON'S BAY COMPANY

Stromness became a huge source of labour for the Hudson's Bay Company when recruitment began in 1702. By 1799 Orcadians were at the forefront of the enterprise, with 416 of the 530 men working for the company coming from the islands. The company is said to have preferred Orcadians as they were 'more sober than the Irish and cheaper than the English'.

paid for by a private individual or company, which operated alongside official naval vessels, but was able to operate outside the rules. One of these ships was the *Liberty* under Captain Scott, which was responsible for attacking British merchant shipping all across the Atlantic and into British coastal waters. It was in this regard that the *Liberty* found herself off the coast of Scotland in 1813, where she came into difficulties at the entrance to Campbeltown harbour and ran aground.

A cannon from the ship found its way to Stromness where it was put to good use in the service of Canada, being fired to announce the arrival of Hudson's Bay Company ships into the harbour.

5 Walking out of town and past the golf course you soon find yourself looking at a scene reminiscent of a war movie. This is **Ness Battery**, a series of defensive positions installed over two world wars, which give an eerie feel to this stretch of the walk.

With the emergence of Germany as a naval power at the turn of the 20th century, the threat to British marine interests was judged to have moved from the Channel to the North Sea and the Atlantic. The decision was taken to relocate the fleet to an anchorage both large enough to take the ships and allow easy access to whichever part of the British Empire required their services.

Scapa Flow was chosen and on the eve of the First World War the fleet moved north to Ness, where fortifications were built to defend Hoy Sound, the main western approach into the bay.

With the signing of the Armistice in 1919, the German fleet was ordered to sail to Scapa Flow, where it was interned awaiting the decision of the Versailles peace conference as to its fate. On 21 June, fearing that the peace treaty would deliver his ships to the British, the German commander, Rear Admiral von Reuter, ordered the scuttling of his fleet. Fifty-one of the German vessels went to the bottom, with a further eight being beached.

The defences were further increased with the outbreak of war in 1939 and most of what you see today dates from this period.

6 Further round the shore from Ness Battery you will come to a cemetery jutting out at the cliff's edge. **Warbeth** (or Warebeth) **cemetery**, appropriately within touching distance of the sea and boasting beautiful views over to the hills of the island of Hoy, is the resting place of most citizens of Stromness. Within the cemetery, alongside the provosts and air force pilots and sailors who lie here, is the grave of George Mackay Brown, who died at his house in April 1996.

7 Just past the cemetery, the path takes you along the edge of a small cliff that skirts round **Warbeth Bay**. It's well worth stepping off the path at this point and heading down to the beach, which has an international reputation for fish fossils. You will also see ots of seaweed in the shallows just offshore. For the crofters and fishermen who made up the population round Warbeth in the 18th

and 19th centuries, harvesting and processing kelp, in particular, was big business.

8 Heading up the hill from Warbeth Cemetery you will meet the road running across in front of you. Turn right and you will be on the home straight. From this high point there are wonderful **views of Scapa Flow** and it is easy to imagine it a hundred years ago, full of massive warships and their crews sitting at anchor in the peaceful waters of the bay.

The Royal Navy abandoned Scapa Flow in 1956 and the shipping to be seen here today is of a more peaceful nature, mainly oil tankers stopping at the Flotta oil terminal. The gas flare at the terminal can be seen in the distance to the south. Although most of the wealth and benefits from the North Sea fossil fuels have headed either to Shetland or Aberdeen, the Flotta terminal is now a vital part of the local economy.

Just before you begin the descent into Stromness, enjoy the spectacle of the St John's Head cliffs on the island of Hoy, at 1136 feet the highest perpendicular sea cliffs in Britain, too steep even for seabirds to nest on them.

Stromness

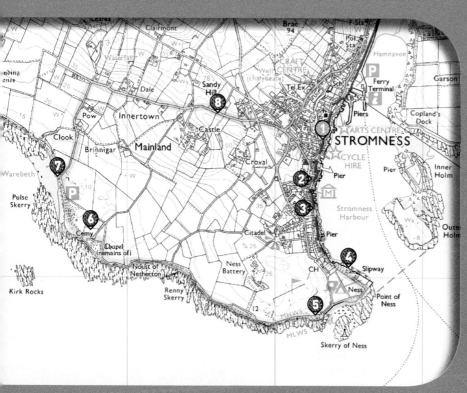

Ordnance Survey Explorer Map number 463
© Crown Copyright 2008

ADVICE
A largely flat walk with good paths at the start. Beyond Ness Point the path does get a bit rougher and the way back to Stromness is mostly uphill.

PARKING
There is a pay-and-display car park at the Ferry Inn or free parking on the street.

START
The town of Stromness is built along one main road, Dundas Street. Begin walking along it at the northern end and you will soon see the Graham House on the left-hand side.

CONTACT DETAILS
Stromness Tourist Information Centre, Ferry Terminal Building, The Pier Head, Stromness, Orkney KW16 3AA
t: 01856 850716

Spey Bay and Moray
Wood, whisky and wildlife

MEDIUM

ACCESS

8½ MILES

4:30

TUGNET ICE HOUSE

In winter, the shallow floodwater on fields adjoining the Spey would freeze solid. The ice was broken into blocks and stored in the ice house until the salmon season, then taken out and packed around the salmon in boxes for onward transportation.

The Spey is the second longest river in Scotland and on this walk around its estuary you should see plenty of birds and marine wildlife, as well as signs of recent human activity.

The Spey rises in the Monadhliath Mountains to the south-west, provides water for whisky distilleries and salmon fishing, and is a Site of Special Scientific Interest (SSSI). In an old salmon fishing station at Spey Bay there is a shop and interpretation centre that provides information about marine mammals and their conservation. Staff at the centre also give guided tours around the Tugnet Ice House, which was used to store winter ice for preserving the locally caught salmon during transport to market. Around the edge of the car park there is a display of sculptures depicting local interests and activities, with the sculpture of an osprey as centrepiece.

There are views across the estuary to Kingston and into the Moray Firth, which has a resident population of about 130 bottlenose dolphins and, in summer, hundreds of grey and common seal. The shingle and mudflats are visited by redshank, whooper swans, greylag and pink-footed geese, and nearly all Britain's breeding goldeneye duck are found along the river.

1 From the south-western corner of the Whale and Dolphin Conservation Society Wildlife Centre car park, follow the signs

for the Moray Coast walk, but keep to the right along the riverbank. There are bushy willow, gorse and alder here and, on the left, a marshy area with prominent common reed and meadowsweet. On the shingle islands in the river are three invasive exotic species: Himalayan balsam, Japanese knotweed and the even more unpleasant **giant hogweed**. Not quite so unwelcome, but still an alien, is the blue nootka lupin. As you walk on upriver, the trees get larger, with silver birch, sycamore and occasional ash, as well as the alder and willow. At ground level there are sedges and ferns, typical of a damp woodland habitat.

② Just before you reach the old railway track, turn right and walk parallel to it until you come to the **bridge**, where you join the railway to cross over the Spey. Notice the way the bridge, which opened in 1885, is constructed. It stands on granite piers and all the steelwork is joined together with rivets. On the eastern end of the bridge are the remains of poles with numerous crosspieces that in

the heyday of the railway would have carried dozens of telegraph wires for communicating between stations. In the river below you can see the banks of shingle washed down from the hills above. These are constantly shifting, depending on the flow of water.

Continue along the track bed until you reach a stone bridge carrying the road over the railway. Then climb up the steps on the left to join the road. Turn right into Church Road, follow this along past the converted church on the left, then go right by the Garmouth Hotel into South Road. Notice the carefully squared ashlar construction of the house on the right opposite the village shop, a sign of earlier prosperity associated with the timber, fishing and shipbuilding industries, and a plaque marking the site of a 400-year-old traditional market fair.

③ Before the car park and household recycling site on the left, take the path between two white houses and go up the hill to the **water tower** and the Four Poster standing stones, which date back to 1500 BC. There is a fine view over the Spey mudflats and beyond from this point. Carry on along the path, bear right past farm buildings into Burnside Road, then branch left across the marshy area on a narrower path. Cross over the footbridge, where in summer there are water crowfoot and floating sweetgrass in the water below, and musk on the banks. Follow the signs to Kingston, then turn left into the road towards Lein.

④ Go through the car park and recreation ground onto the shingle, from where you have a good view along the coast to east and west. In earlier years, the village was famous for building wooden ships using timber from the pinewoods inland. The ships were built in

SHIPBUILDING

From the 16th century onwards, timber felled in the higher parts of the valley was floated down the river from Strathspey to the coast. They were guided by men in lightweight, skin-covered coracles, supplemented by men on the banks, who pushed off logs that got stuck. The wood was initially destined for export, but it made sense to establish a shipbuilding industry here and over the years some seven shipyards in Garmouth and Kingston built about 700 schooners, including clipper ships to service the tea industry in India and the 800-ton *Lord Macduff*. Spey-built was a mark of quality.

DALLACHY AIRFIELD

Dallachy Airfield was built in 1943, initially for training purposes. During 1944–1945, Beaufighter aircraft of the Dallachy Strike Wing played a major role in disrupting supplies to the Third Reich by attacking shipping in the North Sea and off Norway. However, some 70 aircrew lost their lives in these attacks.

THE ALIENS HAVE LANDED

Species such as Himalayan balsam and Japanese knotweed were originally introduced to Britain as garden plants, and then escaped into the wild. The balsam has exploding pods that scatter its seeds widely and the knotweed can regenerate from the tiniest fragment of root. In an environment without their normal grazers, pests or diseases, they can outcompete and replace native species.

the open on the shore, so all that's left of this industry now are workers' houses, many of them built from a mixture of rounded pebbles and clay, weatherproofed with cement harling, and some **larger properties belonging to the yard owners**. Return to the recreation field and turn left along the track, past the bus stop and the red telephone box.

5 In the car park ahead, once the site of the main shipbuilding yard, there is a **stone commemorating the bicentenary of the village**. Follow the road back from here to Garmouth, past saltmarsh on the estuary side. The channels through this have shifted many times and were once supplemented by channels dug through to the estuary to float timber and boats in and out of the shipbuilding yards. Carry on past the golf course, taking care where the road has no footpath.

Rejoin the earlier route beyond the car park, noting the stone in the wall of the large house on the left commemorating the landing of the future King Charles II and his covenant with the local people. Retrace the route to the road bridge and rejoin the Moray Coast path back over the railway viaduct. Then keep straight on along the track bed until you reach the B-road.

6 Turn right. Take care on the stretch without a footpath in front of the Spey Bay public hall, then turn left along the road for Nether Dallachy.

As you enter the village, you climb slightly up the remains of an old river bank. Keep straight ahead at the T-junction, following the path into the pinewoods, past various concrete structures associated with the Second World War Dallachy Airfield.

7 Coming out onto the road again, the derelict control tower stands in the middle of the field to the south. Turn left along the road, through the remains of the old diamond-shaped hardstandings used as dispersal points for aircraft. In the wood to your left is an old gravel pit, which is now being used for landfill.

After about 100 yards, bear left along the track under the electricity line and follow

this up into Lower Auchenreath. Keep to the right, then go left at a T-junction, passing the **field full of arks** for outdoor pigs on the right of the road. Bear to the left of the farm buildings and head along the track to the old railway bridge.

8 In the trees on the left, the course of the railway disappears into a flooded area of old gravel workings and from the railway bridge there is a clear view over the **salt marsh** to Buckie.

Ignore the signs for the coastal path on the left and keep straight on to the golf course. Taking note of the warning notice about flying golfballs, head over to the shingle bank near the green 150-yard marker. Go left along the edge of the bank, past stands of gorse, heather and broom. On a clear day you may be able to see the hills to the north of the Moray Firth from the top of the shingle bank.

Follow the line of the shingle to the west, either on the grass or on the shore, depending on the tide. Note the wide range of different pebble colours and the size of the stones thrown up onto the shingle bank. You should also keep a lookout for different gull

species and you may see osprey, dolphins and other marine mammals. In summer, the vegetation includes thrift, haresfoot clover and bugloss, along with various grasses, including marram. After just over a mile, pass to the seaward side of the Spey Bay Hotel and other houses, noting the turf roof on one of the sheds, and return to the Wildlife Centre.

Spey Bay and Moray

ADVICE

Much of the walk is on level, well surfaced tracks and can be undertaken with a wheelchair or buggy. The route forms a figure of eight and, if preferred, could be taken as two separate walks of 4 and 4½ miles, taking 2 and 2½ hours respectively. Dogs should be kept on the lead when on the roads. There is a cafe and toilets at the Wildlife Centre and a shop and pub in Garmouth.

PARKING

There is ample parking at the Spey Bay Centre or in the surrounding roads, but the centre is not accessible by public transport. However, there are buses to Garmouth and Kingston, which lie on the route, and the walk could be started from either of these places.

START

The walk starts and ends at the Whale and Dolphin Conservation Society's Spey Bay Centre.

CONTACT DETAILS

Whale and Dolphin Conservation Society Wildlife Centre, Spey Bay, Moray IV32 7PJ
t: 01343 820339
f: 01343 829065
e: wildlifecentre@wdcs.org
w: wdcs.org/wildlifecentre

Ordnance Survey Explorer Map number 424
© Crown Copyright 2008

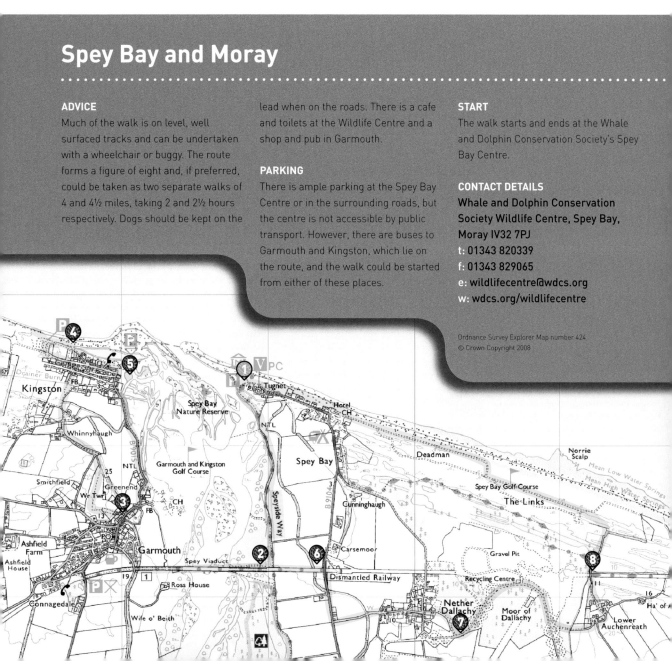

Aberdeen
Spectacular scenery around the Granite City

EASY

ACCESS

3 MILES

2:00

①

This walk offers striking views of the Granite City, including Aberdeen docks, but it also takes in interesting wildlife and some stunning cliffs and seascapes.

① The **Bay of Nigg** was probably named after the Norman Cormac de Nug, who had a stronghold in Torry. At one time the Dee flowed through the Vale of Tullos into this bay, but it was diverted to its current course during the ice ages. The bay used to be a major recreational destination for the inhabitants of Aberdeen, but is now relatively quiet, apart from lunchtime walkers and surfing enthusiasts.

② Leave the north-east corner of the car park following the obvious path past the derelict fishing station. Wheelchair users can follow the road, taking care with traffic.

The concrete on the seaward side of the path marks the line of an old outfall from the city drainage system, via the abandoned valve house below the lighthouse.

A steep climb takes you up to the grade A-listed Torry Coo, the **Girdle Ness foghorn**, whose melancholy voice warned ships of the rocky shelf extending out from the shore below. Improved navigation aids mean it has been silent since the 1980s, but it was recently saved from demolition and partially restored.

The 128-foot high lighthouse, designed by Robert Stevenson, grandfather of the *Treasure Island* author Robert Louis Stevenson, was

built in 1830. It is now fully automated, but when it was staffed, the coastguards were allowed to graze their livestock on Walker Park to the south-east of the lighthouse. The need for the lighthouse was emphasised in 1813, when the whaler *Oscar* sank on the rocks of Greyhope Bay, to the north of Girdle Ness, with the loss of 55 lives.

3 From the **lighthouse**, follow the pavement along the roadside, around the shore of Greyhope Bay. Look out for birds, as the Girdle Ness headland is a haven for both migrant and resident species.

Leave the road by the track leading to the South Breakwater and bear left along the coastal path. The concrete breakwater, which is not accessible to the public, was built in the early 1870s to supplement the older Inner South Breakwater. The navigation light in the tower on the end of the breakwater helps to guide vessels into the harbour.

4 To the west of the breakwater there is a small sandy beach, at the top of which is a stand of **marram grass**. This has a key role in stabilising dunes through its extensive underground rhizome system. This is especially important to the north of the harbour where there is a 12-mile stretch of sand, one of the longest stretches of beach and dune coastline in Scotland and a major resource for wildlife and recreation. Another characteristic species is the sea plantain, but a less welcome plant on this section of the walk is the invasive alien Japanese knotweed.

Among the shingle there are piles of mussel shells and there is a larger mussel bed on the far side of the North Pier opposite. The pier is some half-a-mile long, with a lighthouse on its seaward end. It was extended in three stages during the 18th and 19th centuries to help combat silting up of the harbour by sand brought by waves from the beaches to the north, and to ward ships off the rocks in the water along the headland. On the south shore the next landmark is the Inner South Breakwater, which was completed in 1840 following a suggestion by Thomas Telford. It protected the navigation channel from easterly storms and, by concentrating the river and tidal flows through the channel, helped keep it clear of silt.

5 Opposite the shoreward end of the breakwater, take the path leading up across the road to the Torry Battery on the hill, past an interpretation board showing marine life such as the bottlenose dolphins and porpoises that are common visitors to the harbour. The Battery itself, which was built in 1861, is now largely in ruins, but the arched gateway is still intact and on the east side you can still see the **mounting rings** for one of its 60-pounder guns.

From the Battery you can see the contrast between the rocky headland of Girdle Ness and the beaches of Aberdeen Bay to the north. This reflects the underlying geology. The granite quarried from the Hill of Rubislaw west of

the city was shipped through the harbour, along with salmon, grain and livestock from inland and imported timber. Whaling and shipbuilding also brought prosperity at different times and, most recently, geological formations offshore have yielded North Sea oil.

6 Rejoin the coast path below the Battery. From here you can see the **Marine Operations Centre**, a major landmark on the north pier. The centre, which cost over £2½ million, opened in 2006 and each year oversees 25,000 vessel movements in and around the harbour. It superseded the earlier Navigation Control Centre in the octagonal Roundhouse some 90 yards further west.

Between these two is a brick obelisk called Scarty's Monument, after the nickname of a 19th century harbour pilot, but it's actually the vent for a disused drainage outfall. A reminder of the earlier history of the port is provided by the hand-operated capstan on the landward end of the small 18th century South Breakwater, which worked a ferry that went across from Torry to the north shore.

Proceeding west along the coast path, directly ahead on the Mearns Quay are the mud storage tanks of various oil companies. This mud is actually a complex lubricating and cooling compound used in drilling for oil. There is a major display of maritime history and technology in the Maritime Museum on Shiprow in the city.

Continue along the coast path until it joins Greyhope Road, opposite the allotments, and cross with care into St Fitticks Road on the south side. Walk up this road, past the

Marine Laboratory and the Nigg Bay Golf Club clubhouse (if you come to the area by bus, you start and finish the walk at this point).

The ruined church of St Fittick lies in the open area on the right of the road and is worth a look. Then walk across the recreation area towards the low buildings of the wastewater treatment plant at the south end of Nigg Bay. Recent developments such as this have improved coastal water quality – it was once 'unsatisfactory', but is now 'good'. Join the metalled path, then turn left along the grass towards the road and right across the burn out onto the road.

7 Cross over the road with care, clamber over a small embankment that blocks the entrance to a disused section of road, then follow the path up the hill parallel to the road. To your right, across the other side of the road and the railway cutting, which opened in 1850, the land rises towards the Hill of Tullos. Until recently this area was used as landfill for the city's waste, but it is now being restored. At intervals, you will see the tops of **boreholes** used to monitor the state of the landfill material and groundwater.

The path now runs along the top of an eroding cliff of Quaternary clay and gravel beds. During the summer, this area is a mass of wild flowers, including gorse, hawkbit, heather, spear thistle, harebells and bird's foot trefoil. The infertile soil restricts the growth of the faster growing grasses and

similar species that could otherwise crowd out these lower growing flowers.

8 At this point, you can either retrace your route back down to shore level or continue with care along the cliff path, past a series of rocky inlets, culminating in the spectacular cliffs of the **Needles Eye**. Care is needed all along this section. After about 800 yards, a path to the right emerges onto the road near a bridge over the railway. You can either walk back along the road past the coastguard radio station or retrace your steps along the path. Finally, walk back across the gravel beach to the car park, noting the variety of pebbles and coarse sand underfoot.

NORTH SEA OIL

Oil forms from the covered remains of creatures that lived in the precursor to the North Sea around 160 million years ago. Pressure and heat converted the organisms into oil that was trapped in reservoirs of porous rock, under caps of impermeable material.

Aberdeen

ADVICE

Most of this walk is on open land or roads and is easily accessible. Wheelchair and buggy users can follow a similar circular route entirely on tarmac. There are no toilet facilities or refreshments anywhere on this walk, but plenty in Aberdeen itself, which is less than a mile away. A southward cliff-top extension of the main circular walk is not suitable for wheelchairs or buggies and close supervision of young children would be necessary. The only livestock are in a field adjoining the southern extension of the walk, but dogs should be kept under control nonetheless.

PARKING

From Aberdeen, follow signs to Torry and park in the free car parks at Nigg Bay or Torry Battery. There is also a bus service from the centre of Aberdeen, which stops by the entrance to the Nigg Bay Golf Club on the Balnagask course.

START

The walk starts and finishes in the car park at Nigg Bay.

CONTACT DETAILS

Aberdeen Visitor Information Centre, 23 Union Street, Aberdeen AB11 5PB
t: 01224 288828
e: aberdeen.information@visit scotland.com
w: visitscotland.com

Ordnance Survey Explorer Map number 406
© Crown Copyright 2008

Perth and Tayside
The 'singing sands' of Lunan Bay

MEDIUM

ACCESS

4 MILES

3:00

With odd-shaped rocks in the north and relics of the Second World War in the south, the beautiful 'singing sands' of Lunan Bay make up one of the loveliest walks in Scotland.

Three miles south of Montrose you will find yourself at Boddin Point, a low-lying spur of rock that marks the northern end of Lunan Bay. The car park has fine views south across the bay. Walk to the top of the cliff where you will find a tiny and unremarkable cemetery.

Here is buried **George James Ramsay**, not a notable man in life, he has become a tourist attraction in death. Take a close look at the inscriptions on each tomb. When you get to George James Ramsay's, you will find yourself presented with an extraordinary fact – George was born on 24 November 1859 and died on 17 December 1840.

Was this an error by a careless engraver and not picked up by an illiterate grieving family? Or is the explanation that the relatives knew the error had been made, but could not afford to have the inscription

changed? No one will ever know, but George James Ramsay, thanks to his tombstone, will never be forgotten.

2 The sea carves its mark on land in many ways and just below the cemetery there is an example that has become a minor tourist attraction in its own right. The rocky protuberance that juts out to the sea with almost sheer sides has been noted by visitors for many years due to its curious shape, which becomes more and more apparent as you walk round the corner back towards Boddin Point.

The outcrop was more properly known as the Rock of St Skeagh, but today's visitors know it as **Elephant Rock**, with the shape of the animal clearly defined, its trunk almost dipping into the sea. Erosion has clearly been the culprit, with the tide carving two arches in the protruding rock over the years, which can even be walked through at low tide.

3 Down the hill from the Elephant Rock you will come to the remains of a **salmon fishing station**. It is unclear when fishing for salmon off the Angus coast began, although it is documented that monks were using stake nets here in the Middle Ages. The industry really took off in the early 19th century when Boddin fishing station was built. The station comprised of an icehouse, fish-house, harbour and bothy. The building at the shore is most likely to have been the fish-house.

Nets in use today can be seen up and down the coast for several miles from here. Some nets sit offshore, with others sitting tethered on dry land for repair. Both fixed and sweeping nets are used to catch the fish as they swim along the shoreline. During the early part of the last century, this stretch of coast was the most productive salmon fishery in the British Isles.

After a long period of declining numbers of salmon, fishing ceased here during the 1980s. However, the industry has had a new lease of life since then, with the former licence holder, J. Johnston & Sons of Montrose, selling the rights to independent fishermen who have started their own successful salmon fishery in the bay.

4 The focus of Boddin Point is an **18th-century limekiln** that stands guard where the rocks meet the sea. Like many others up and down this coast, the kiln was built around 1750 at a time when much of the country's landowning class was engaged in 'land improvement', in which lime was used to reduce acidity in the soil.

The kiln was built by local landowner, Robert Scott, of nearby Dunninald Castle, on the discovery of a rich seam of limestone. The idea was that ships could call into the natural harbour at Boddin Point to deliver coal from Fife to burn the limestone and then take away the residue for use elsewhere.

Today the kiln occupies a precarious position on the edge of the rock. Erosion has taken its toll on the seaward side of the

BELL ROCK LIGHTHOUSE

Nearly 200 years after it was first built, the Bell Rock lighthouse still stands proudly flashing its warning light. Eleven miles out to sea off the east coast of Scotland, it is a remarkable sight – a white stone tower over 100 feet high, rising seemingly without support out of the North Sea. In fact, it is precariously poised on a treacherous sandstone reef, which, except at low tides, lies submerged just beneath the waves.

THE BIRDS OF LUNAN BAY

Lunan Bay is a fantastic place for birdwatching. The beach is noted for great grey shrikes and red-backed shrikes throughout the summer months. Other birds spotted here include scoters, rough-legged buzzards and hoopoes, unusual this far north.

KEEP AN EYE OUT

For some, the most valuable natural feature of Lunan Bay is that the beach is well known for its semi-precious stones and agates. The best time to look for them is after a storm, although they can be hard to spot amongst the quartz and other pebbles on the beach.

building and much has already fallen into the sea. Although it is still possible to climb to the top of the kiln for some amazing views along the coast and out to the Bell Rock lighthouse, be aware that the kiln is in serious danger of collapsing into the sea.

5 Turn inland and walk up the hill towards **Dunninald Castle**. The castle sits in beautiful gardens, which were laid out in 1740, and both are open to the public.

A castle was first built here in the 15th century. Black Jack's Castle, as it was known, was built as a fortified house for the Gray family, the same family who went on to sack Red Castle the following century. The family ordered the building of a second castle here after they had been restored to their lands following a period of exile for their part in the Red Castle affair. The next owners, the Leightons, were also no strangers to controversy. Patrick Leighton was accused of robbery by another Montrose merchant after a business deal turned sour.

The castle passed through several owners until the early 19th century, when new owner Peter Arkley commissioned the architect James Gillespie Graham to build the present one. Graham is known as the leading light of the Scottish Gothic style of architecture.

6 Heading southwards along the cliff-top you will see some of the most beautiful beaches in Britain, the stretch of '**singing sands**' between Boddin Point in the north and Lang Craig in the south. The sands are so-called because of the noise you make when you walk on them.

Down on the beach you can see why Lunan Bay was voted best beach in Scotland in 2000. It has been attracting visitors since marauding Vikings landed here in 1010. Visitors these days tend to be a bit more genteel, as they take in the natural beauty of the area or indulge in other popular beach activities.

7 As you walk along the sands your eye is drawn to the brooding ruin on the hill overlooking the beach. This is the famous **Red Castle**, built on the orders of King William 'the Lion' in the late 12th century to deter invaders.

In and out of royal hands, including those of Robert the Bruce, the castle was first called 'rubeum castrum', or Red Castle, in deeds of 1286. This referred to the red sandstone of which it was built. The castle remained a prominent feature of the area until 1579 when a dispute arose after James Gray had married the owner of the castle, Lady Elizabeth Beaton, then fallen in love with her daughter. Lady Beaton threw him out, whereupon James enlisted the help of his brother Andrew and spent the next two years laying siege to the castle.

Partially roofed until 1770 the castle is badly eroded and best viewed from a distance, from where a wide range of birds that use it for nesting can also be spotted.

8 In the spring and summer of 1940 Britain resembled an army camp.

Everywhere fears of German attack were rife, even in sleepy Lunan Bay. Although the major fear was a cross-channel invasion, significant German forces were positioned in Norway. It was not inconceivable that the Nazis could launch raids on the Scottish mainland from there, or even worse a two-pronged invasion, so while the bulk of the British Army was engaged against cross-channel invaders, German forces could land on the weakly defended east coast of Scotland. The decision was taken to build up east coast defences and the Polish Army was given the job, taking up posts from the borders to the northern coasts.

Here at Lunan Bay, in the shadow of a castle built to withstand Viking invaders, lie higgledy-piggledy **defences built in 1940** in the form of huge concrete blocks. These blocks served as anti-tank measures, forming obstacles too steep for German Panzer tanks to negotiate.

To get back to your car, it's best not to rely on public transport as there is little in the area. You are best advised to amble back along the sands and then take the cliff-top path back to the car park at Boddin Point.

Perth and Tayside

ADVICE
This walk is fairly level, involving mostly paved surfaces and flat unmade paths, with sand at the end.

PARKING
Limited parking is available on the track near the cemetery, but there is a car park at Boddin Point.

START
You start the walk in the cliff-top cemetery above Boddin Point.

CONTACT DETAILS
Montrose Tourist Information Centre, Bridge Street, Montrose, Angus DD10 8AB
t: 01674 672000
e: visit@angusahead.com
w: angusahead.com

Dundee
Industrial heritage tour

MEDIUM

ACCESS

6½ MILES

3:30

RRS *DISCOVERY*

Royal Research Ship *Discovery* was launched in Dundee in 1901 as the first purpose-built research vessel. Propelled by steam and sail, and designed to withstand being trapped in ice, it carried Captain Scott's first expedition to Antarctica. In 1986 it returned to Dundee, where it forms the centrepiece of Discovery Point Antarctic Museum.

The docks, two public parks and a range of buildings that reflect various periods of Dundee's industrial and commercial prosperity can all be seen on this fascinating city walk.

The port of Dundee grew in the 14th century, declined during wars in succeeding centuries, but prospered in the 18th century, mainly due to linen production. Jam and jute, along with shipbuilding and whaling, replaced linen in the 19th century, but these industries declined in the 20th century, as publishing, new technology and tourism grew.

From its entrance, walk around **Discovery Point** and along the shore past the back of the Olympia Leisure Centre. There is a good view of the Tay road bridge here. Turn left at the obelisk, then right along the cycleway under the approach road to the bridge and along Victoria Dock Road. Cross the road to the Apex Hotel and turn left into West Victoria Dock Road. The redundant North Carr lightship and HM Frigate *Unicorn*, the oldest British warship still afloat, are moored in the dock. Most of the dockside buildings have been converted for retail or residential use, but the railway lines used to move cargo around the docks are still visible among the cobbles.

Cross the dual carriageway and go left past the imposing stone buildings, cross the cobbled Gellatly Street and turn right

up Commercial Street. Notice the statue of Admiral Duncan on the left just into High Street. Bear left into Meadowside in front of the Gothic Revival-style McManus' Art Gallery and Museum, whose exhibits chronicle Dundee's colourful past as a port. Along on the left is the Howff, an historic graveyard that was the meeting place for the Nine Trades of Dundee described on the interpretation panel. James Chalmers, inventor of the adhesive postage stamp, is buried here. Note several umbrella-shaped specimens of the local Camperdown elm.

2 Cross over Meadowside, go around the *Courier* newspaper building and head towards the neoclassical front of the High School. Turn left into Euclid Street, then right into Constitution Road, passing the striking pink sandstone buildings of Abertay University. Bear left into the dual carriageway underpass and up the steps on the right (or loop round to the left and back along the road). Carry on up Constitution Road to the junction with Barrack Road. Bear right into Laurel Bank and Prospect Place, where the large Victorian houses with their tree-filled gardens reflect Dundee's commercial prosperity in that period.

Return to the junction with Barrack Road and go right, past the front of Dundee College. Keeping to the left of the road gives good views over an area of the town that was once a major site for jute manufacture. There is a display dealing with this period

in the Verdant Works mill, whose chimney is visible from this point. Bear slightly left into **Dudhope Castle Park**. The castle itself was built in the 13th century to replace an earlier Dundee Castle and then rebuilt in the 16th century. From the north-west corner of the car park go up the steps then right, back onto the road opposite Smillie Court and the former hospital buildings (wheelchair and buggy users could return to Barrack Road and use the pavement up the hill).

3 Turn left along Dudhope Terrace, then right into Law Street. Take care, as the street is steep with an uneven pavement surface. Go right at Kinghorne Road, left into Law Road and then left again following the brown tourist sign to Dundee Law. There are good views over the Tay road and rail bridges from higher up this road. Carry on past the allotments to the car park and public conveniences.

At this point, you can either continue along the tarmac road that spirals up to the **war memorial** on top of the Law or take the very steep short-cut path on the right. From the Law, there is a panorama of the city, across the Firth of Tay to Fife and the surrounding countryside. The city's importance as a port and manufacturing centre shows clearly in the docks and in the mill buildings such as Cox's Stack, the prominent chimney to the north-west.

THE MILLS OBSERVATORY

The Mills Observatory was built in 1935 with funds from John Mills, a local linen and twine manufacturer and keen amateur scientist. It was specifically for use by the general public and houses various scientific instruments, including a 10-inch Victorian refracting telescope and a small planetarium.

SAND AND LAVA

Dundee lies on Devonian sandstone from around 400 million years ago, which has been quarried extensively for local buildings. Around 370 million years ago, hot molten rock was forced up from below and the lava solidified into hard rocky plugs. These resisted subsequent erosion by water and ice, leaving the protruding Law and Balgay hills.

4 The descent from here to the west is very steep and includes several sets of steps, but by returning along the road it is possible to get a bus back to the start and then a second bus to the next landmark, Balgay Park. Walkers should go down the path to the west, through the belt of Scots pines and more allotments to Lochee Road. Turn right past Balgay Church into Tullideph Road, with the Franciscan Friary on the left. Turn left at the traffic lights into City Road, then right along Saggar Street at the sign to Balgay Park. Carry on through the houses onto the steep grassy path straight ahead.

The hill of Balgay Park was originally part of the estate of Sir William and Lady Scott. Some of the wide range of mature trees in the park date from their ownership, with others being planted after the area was bought by the Council and opened to the public in 1870. At the end of the path, among the trees, there is a short flight of steps down to a tarmac track, where you turn right then follow the sign on the left to the **Mills Observatory**, the only public observatory in the UK. Take the signposted Planet Trail.

5 Walk to the right of the observatory through the beech trees, circling down the hill to an ornate 1877 **cast-iron bridge** across the sunken road. Continue along the path, down through the trees to the gazebo on the right of the road. Keep left along the road, past the open ground of Victoria Park on the right, and past rhododendrons and more mature specimen trees for which the park is justly famous.

6 At the entrance to the park, where buses from the city centre stop, there is a good view back to Dundee Law. Turn right down the tree-lined Balgay Road and cross Blackness Road into Blackness Avenue, where you get occasional views between the houses to the Tay rail bridge. Note the fine sandstone terrace of houses on Seymour Street to the left. These show interesting architectural detailing, including ornate chimneys, balconies and stone window reveals. The

stone-built tenement housing for mill workers further down on the left is plainer and much more crowded, but still impressive. Notice the contrast between the carefully squared sandstone ashlar on the fronts of the houses and the uncut rough stone on less visible end walls. At the junction with Hawk Hill and Perth Road is the monumental Blackness Library, one of many across Scotland funded by Andrew Carnegie.

7 At this point, cross over the two roads to Shepherd's Loan (loan means lane), where there is another large **mill building** with ornate windows. Past the mill, you come to another grassy area. Skirt along the edge of this to the end, past the redundant Dundee

Rope Works building on the left. Cross over the road to the car park opposite and join the pedestrian and cycle path in its south-east corner. Carry on along this, back towards Discovery Point.

The land you're walking on was reclaimed from the sea in earlier centuries as the harbour was extended out into deeper waters. It is about to be redeveloped and the changing history of Dundee is well illustrated by the mix of ultra modern and earlier architecture visible on its inland slopes.

8 Continue past the Sensation Science Centre, cross the road at the traffic lights and walk round the HM Inspectorate of Education building under the truncated overhead walkway to the front of the station. This walkway used to join the station to the town centre, but will disappear when this area is redeveloped. Cross the dual carriageway to return to Discovery Point.

Dundee

ADVICE

The walk involves some steep climbs, with steps and some surfaces unsuitable for wheelchairs or buggies. However, the main points on the walk can be reached via alternative flatter routes. Dogs should be kept on a lead at all times.

PARKING

There are a number of pay-and-display car parks near the start of the walk.

START

The walk starts and ends at Discovery Point Antarctic Museum, adjacent to Dundee railway station.

CONTACT DETAILS

**Dundee Tourist Information Centre,
21 Castle Street, Dundee, Angus DD1 3AA**
t: 01382 527527
e: dundee@visitscotland.com
w: angusanddundee.co.uk

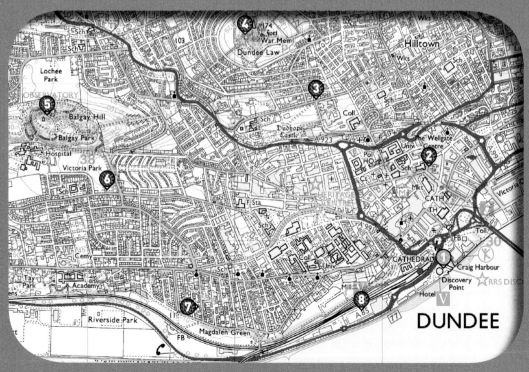

Ordnance Survey Explorer Map number 380
© Crown Copyright 2008

Old Leith
On the history trail

MEDIUM

ACCESS

3 MILES

2:00

Now part of Edinburgh, the town of Leith has its own ancient and fascinating history. Take a stroll through some of its key moments on this coastal city-walk.

❶ The walk starts by the narwhal monument at Kirkgate shopping centre in the heart of new Leith, but hidden among the modern amenities are traces of the town's ancient history.

Leith has been a strategic military point at many times through Scottish history. It is proud of its reputation as a Scottish stronghold where kings and queens would rally their armies. In July 1650 Oliver Cromwell tried to break through the fortifications organised by John Mylne, the King's Master Mason, and defended by a Covenanting army under Sir David Leslie.

Historians have argued that the Scots army should have been satisfied with this successful defence as Cromwell turned away and was leading his troops in orderly retreat. But Leslie decided to chase him to Dunbar, where the Scots met resounding defeat.

Before moving off from this spot notice the disused railway station opposite, which was featured in the 1996 film of Irvine Welsh's novel, *Trainspotting*.

❷ Head away from Leith Walk and along Kirkgate. After 50 yards you'll come to **South Leith Parish Church** on your right and Trinity House on the left. Trinity

House was founded in 1555 as a hospital for mariners. It was funded by 'Prime Gilt', a tax levied on every ton of cargo that passed through the town. For seafarers who knew their journeys were fraught with danger, it was a kind of insurance policy for their families back home.

The first building was replaced in 1816 by the one you see now, but the cellars from the original hospital remain and have had many uses over the years. Leith is famous for the import of wine and it is almost certain that the finest claret and burgundy would have been stored here. In 1636 it was used as a grammar school and around the same time Oliver Cromwell used the cellars as a store for his army. There is still a section of the original building round the corner in St Anthony's Place where you can graffiti from 1555. Today you can visit Trinity House museum.

South Leith Parish Church sits on an ancient site. A Templar hospice was founded here in 1128 where, according to records, Robert the Bruce was treated for leprosy. William Wallace is said to have written his famous letter to the burgers of Marleburg from here to let them know it was safe to return to Scotland to trade. The church, which replaced the hospice around 1390, was looted by the English in 1544 during the invasion known as 'the Rough Wooing'.

❸ Walk through the graveyard and you will come to a gate on the far side. Go through the gate and over Constitution Street into Links Lane. You are now standing on the corner of **Leith Links** – the real home of golf.

In the 15th century the links were used for a number of things: mustering troops, archery practice, exercising horses, grazing cattle, fairs, breeding rabbits and then, in 1457, came the first mention of golf. The game did not initially receive royal approval. King James II called it 'utterly cryt downe and nocht usyt' as it interfered with archery practice, but by 1505 James IV was playing it.

In March 1744 the Honourable Company of Edinburgh Golfers was formed here. They published the first rules of the game and held their first competition. Played over five holes, each over 400 yards, they used feathery balls and contested for a silver club presented by the City of Edinburgh.

❹ From Leith Links retrace your steps to John's Place, turn right down towards the docks then down to the junction and go left along Queen Charlotte Street. At the end take the right fork and head down Water Street. Turn left into Burgess Street where you will find Lamb's House.

Mary Queen of Scots landed at **Leith harbour** on 19 August 1561 after returning from France. Because she was still refusing to sign the Treaty of Edinburgh, Elizabeth denied her cousin passage through England and Mary had bravely sailed the whole way from Calais to Leith.

It is reported that she had arrived earlier than expected and that preparations had not been made for her arrival at Holyrood. She and her retinue were taken to dine and stay the night in Andro Lambis House before continuing to Edinburgh the following day.

LEITH LINKS

On 7 March 1994, the 250th anniversary of the Honorable Company of Edinburgh Golfers was celebrated by the planting of a white hawthorn tree near the finish of the fifth hole. On 19 March a six-a-side two-ball foursome challenge match was contested between the High Constables of the Port of Leith and the Honourable Company. The Honourable Company won by two matches to one.

THE SCOTTISH EXECUTIVE

The Scottish Executive building is on the site of the old Victoria Quay. Opening in 1996 it originally housed parts of the Scottish Office. Since the devolution of Scotland in 1999 it has become home to over 2000 civil servants in the service of the current Scottish government. It is argued that this building kick-started Leith's regeneration programme, which now includes new flats, converted warehouses, bistros, bars and restaurants.

⑤ From Lamb's House carry on down Burgess Street towards the water and turn right at the end onto the Shore. Cross over Bernard Street and a few yards along on your right is a plaque commemorating the **King's Landing**.

In August 1822 George IV landed where you're standing – the first visit from an English monarch for nearly 200 years. Government ministers had pressed the king to bring forward a proposed visit to Scotland to divert him from diplomatic intrigue at the Congress of the Nations in Vienna.

Despite unkind caricatures of the king as 'our fat friend in tights and a kilt' the visit was a great success – there was a marked increase in goodwill towards the crown and between fellow Scots – and the kilt, banned by the government after the Jacobite Rebellion of 1745, was adopted as the national dress of Scotland.

⑥ From the plaque make your way further along the Shore past the abandoned ship on your left and the Shore Bar on your right. Across the courtyard by Malmaison Hotel is a **bench with a statue** of a man sitting on it and a harpoon in the corner, designed in memory of the whaling industry.

Though now you may think that Christian Salvesen is merely a haulage company whose lorries you see on the motorway, it was once the largest whaling company in the world with offices in Leith's Bernard Street.

Originally from Norway, the company dominated business in the town for 100 years.

Though the company was active in many other areas of trading by sea, whaling dominated its profits in the early years of the 20th century. In 1907 the company started Antarctic whaling. By 1914 Salvesen's whaling fleet consisted of 25 vessels. The company ceased whaling in the 1960s, but remains as one of the town's most successful business ventures.

⑦ From the statue, continue on by the walkway through Victoria Bridge, then turn left following the roadway by Ocean Drive opposite the Scottish Office to **Commercial Quay**. This is the heart of the new Leith – busy with innovative restaurants, bars and cafes. But as you stroll along the cobbled pedestrian walkway and soak in the relaxed atmosphere, note that this is the old dock area of the town, once home to the East and the West old docks, and the Victoria dock.

On the south side of the docks, running the full length of them, were bonded warehouses where whisky from the Highlands was bottled and kept until the duty was paid. The golden liquid was then released and sent out to order.

Today, the Scottish Executive building stands on reclaimed ground where the old docks have been filled-in.

⑧ You now have to double-back on yourself along the cobbled pedestrian walkway. Walk back to the crossroads and traffic lights by the corner of the Shore and

Bernard Street (by the abandoned boat) and turn right. Walk up by the side of the water and take the first right onto St Ninians Bridge.

This part of Leith has an unsavoury past. It was known as the area where pirates and rogues hung about, and was even reputed to be haunted. But the most gruesome story of all involves the Barton family and a barrel of pickled heads.

Andrew Barton, a member of a prominent Scottish family of merchants, had earned the favour of King James III with his skills as a commander at sea. In 1476 the Portuguese, then 'rulers' on the high seas, captured and murdered Andrew's father, John. The king granted the family letters of marque – basically written permission

to take revenge on those responsible. For the next few years the Bartons carried on a family war with the Portuguese, further enhancing Andrew's reputation.

When James IV came to the throne a crisis occurred when a group of Scottish merchants were attacked and killed by Dutch pirates. Andrew Barton was sent to seek revenge for the Dutch assault. It's said that he succeeded in clearing the Scottish coasts of the Dutch ships and sent the king a number of barrels full of the heads of the Dutch pirates as proof of his good work.

The walk ends with this grizzly tale. Turn back on yourself again and its an easy walk back to the starting point along Tolbooth Wynd and Kirkgate to the foot of Leith Walk.

Old Leith

ADVICE

Although the walk is on paved walkways for the most part it is longer than it looks. Rest your feet and stop for a cuppa or a pint on the way.

PARKING

There are plenty of pay-and-display car parks in and around Leith. However, if you are staying in Edinburgh buses from the city centre to the foot of Leith Walk include the 16, 11 and 15.

START

The walk starts at the narwhal monument just outside the Kirkgate shopping centre on Leith Walk.

CONTACT DETAILS

e: auldedinburgh@hotmail.com
w: www.auldedinburgh.co.uk

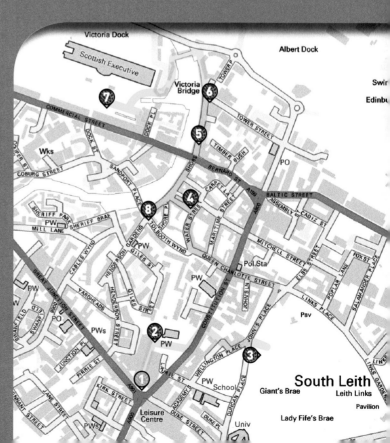

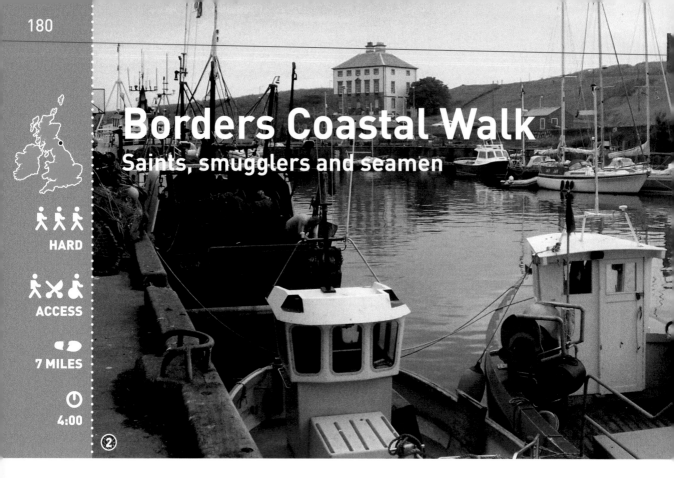

Borders Coastal Walk
Saints, smugglers and seamen

HARD

ACCESS

7 MILES

4:00

②

Taking the coastal path north from Eyemouth, this bracing walk takes you through the villages of St Abbs and Coldingham, and the spectacular nature reserve at St Abbs Head.

❶ The walk starts in the shadow of Gunsgreen House, which dominates Eyemouth from its position above the harbour. A beautiful 18th-century merchant's villa, the house hides some dark secrets within its walls.

It was designed and built by James Adam around 1755. James Adam was the younger brother of the more celebrated Robert Adam. Gunsgreen is unusual in being solely the work of the younger Adam. However, the house is not only notable as a work of architecture, it was also the centre of a lucrative trade – smuggling. Eyemouth was popular with smugglers, being the closest Scottish port to the continent, and throughout the town there were hidden stores and secret passages. Gunsgreen itself was alleged to have its roof space full of illicit items, hiding places built into the walls and, most inventively, a fireplace that swung open revealing a hiding space behind it.

❷ Make your way across to the **harbour** using the walkway by the house. Head for the viewing deck, which allows a great view of the country's second largest inshore fishing port. Fishing has been an important part of the local economy since the Middle Ages. However, the sea has not always been a kind mistress and has also taken back from the town. Its form of currency has been the lives of the fishermen who have ventured out to the treacherous seas beyond the harbour wall.

The biggest single toll came on a peaceful October day in 1881, the day that would become known to the town as Black Friday. It was a morning of glorious sunshine when the Eyemouth fishing fleet set out. There were 45 boats at sea when the sky blackened and the barometers dropped. The storm broke around midday, unleashing hurricane-strength winds. As the boats headed for the safety of the harbour they were dashed against the Hurkar Rocks. Fatalities totalled 189 men and boys. Today, as a mark of respect to those who died in 1881, the fleet never sails on a Friday.

However, the effects of the disaster proved a catalyst for change and the upshot was a rebuilt harbour, which forms the basis of what can be seen today. Cross back over the walkway and walk down the side of the harbour towards the sea, where you will come to the recently refurbished fishmarket.

3 Turn left at the fishmarket and head for the shore, where you will come to a flight of steps. Walk up these steps and you will be standing at **Fort Point**.

This apparently unremarkable promontory at the north end of Eyemouth is in fact the site of a huge leap forward in military engineering and castle design.

This area has been a site of strategic importance since time immemorial – there are remains of an Iron Age fort on the next headland up – its position allowing its occupiers to dominate the area and control the coast road between England and Edinburgh.

The first fortifications here were constructed by English forces in 1547 during the period known as 'the Rough Wooing' and featured the first use in Britain of a revolutionary new technique.

Late in the previous century, Italian engineers had devised a new method of castle building to counter the new threat posed by powerful field artillery. The new design, called

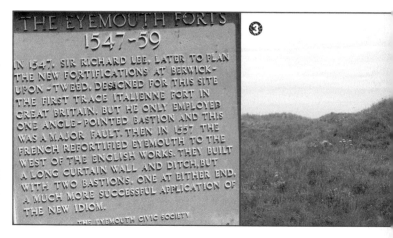

trace italienne, constituted a radical break with the design principles that had governed castle building over the previous centuries. Rather than building high and round, the focus was now on building lower walls, with protruding angular bastions at the corners.

The depth of the new castle's walls provided the main obstacle to incoming cannon fire, especially when supplemented by layers of earth placed in front of the stonework, which absorbed much of the energy of the cannonballs. The walls were also built at oblique angles, which deflected some of the force of the shot, a principle still used today in the design of tanks.

Eyemouth Fort, then, holds the distinction of being the first 'modern' castle in the country, although it only enjoyed a short lifespan, being abandoned and demolished in 1550.

LUCKY PUP

On 17 October 1907 St Abbs mourned the unknown sailors who had lost their lives on the rocks offshore. But the following day one survivor from the shipwreck was spotted – a great dane was found wandering the cliffs, shivering and starving. The dog, named Corra after the rocks off St Abbs on which the *Alfred Erlandsen* had met its fate, became a local celebrity, and, during the First World War, was paraded throughout the area with the tale of its survival told in a bid to raise money for the war effort.

STUNNING VIEWS

For the more adventurous, paths lead up to the lighthouse at the far side of the St Abbs Head nature reserve, with a way back down via the Mire Loch. It's a steep walk, but one which provides for some stunning views of the bird life in the reserve and of the remains of the seventh-century convent of St Ebba.

FLORA AND FAUNA

It is not just our feathered friends which can be enjoyed at St Abbs, of course. The range of flowers is remarkable and the area is well-known for its butterflies. A bit trickier to spot but no less common is the range of fauna on the shoreline and in the waters, including seals, wolf fish, crustaceans and a wide selection of seashells.

Turn northwards along the coastal path. The stretch of coast between Eyemouth and St Abbs provides exhilarating views of the rocks, the sea and the skies.

4 The beach at **Coldingham Bay** could not be more different. You can see why this lovely sandy shore has been a popular tourist destination for many years. More recently Coldingham has become a popular venue for Scotland's burgeoning surfing scene.

5 When the coastal path ends follow the road downhill to **St Abbs harbour**, not forgetting to stop off for a fresh crab roll at the local cafe. A busy fishing port with treacherous rocks offshore, St Abbs was the site of yet another tragedy during a storm on the night of 17 October 1907. While only a short distance from shore the Danish cargo vessel, *Alfred Erlandsen*, headed for Grangemouth

with a cargo of pit props for the Stirlingshire coalfields, was slammed into the Ebbs Carr rocks. As the stricken vessel sounded her horn, the villagers rushed to the harbour to hear the cries from the sailors through dense fog. Lifeboats set sail from Eyemouth and Dunbar, but the situation was hopeless and the lifeboats returned unable to help, and 16 unfortunate seamen lost their lives.

6 Follow the main road out of the village until you find the **St Abbs Head nature reserve**. Perhaps the jewel in the crown in an area of outstanding natural beauty, it was formed from volcanic action

and is now one of the most important sites for flora and fauna in the entire country. The reserve is especially noted for the variety of seabirds which nest there.

7 Retrace your steps to the nature reserve visitor centre. From here follow the road to Coldingham. Carry on through the village until you come to a junction. Look to your left and you will see **Coldingham Priory**.

The first religious establishment recorded here was the nunnery of St Ebba. The community, founded around 642 AD, survived until it was burned down in a fire in 683, shortly after Ebba's death.

Four centuries later a royal feud brought about the founding of another monastery at Coldingham. Edgar, the son of Malcolm III, had been deposed as King of Scotland by his uncle and had gone into exile in England. Bent on revenge, Edgar raised an army of 30,000 soldiers and marched north. On the eve of battle Edgar was visited by a vision of St Cuthbert. After a bloody battle Edgar regained the Scottish throne, a victory foreseen in his vision. In gratitude he established a Benedictine priory on the site at Coldingham in 1098.

The priory became a centre point of the wool trade and a seat of learning on the life of St Ebba. However, in 1560 the Reformation signalled the death-knell of monasticism in the country and the lands belonging to the priory passed to the local landowner.

Oliver Cromwell completed the destruction of the priory after his victory at Dunbar when he used cannons to dislodge some Royalists who were hiding there. A new church was built around the ruins and renovated in the 19th century.

8 Retrace your steps back into the village, from where you can catch the bus back to Eyemouth.

Borders Coastal Walk

ADVICE

The route is mainly a flat, rough path, although there are steps at Fort Point and down to Coldingham Shore. A bus service runs between Eyemouth, St Abbs and Coldingham every hour for those who've had enough walking.

PARKING

There's plenty of free car parking in Eyemouth.

START

Head towards the harbour and you will see Gunsgreen House when you get there.

CONTACT DETAILS

Scottish Borders Tourist Board, Tourist Information Centre, Murray's Green, Jedburgh, Scottish Borders TD8 6BE
t: 01835 863435

Ordnance Survey Explorer Map number 346
© Crown Copyright 2008

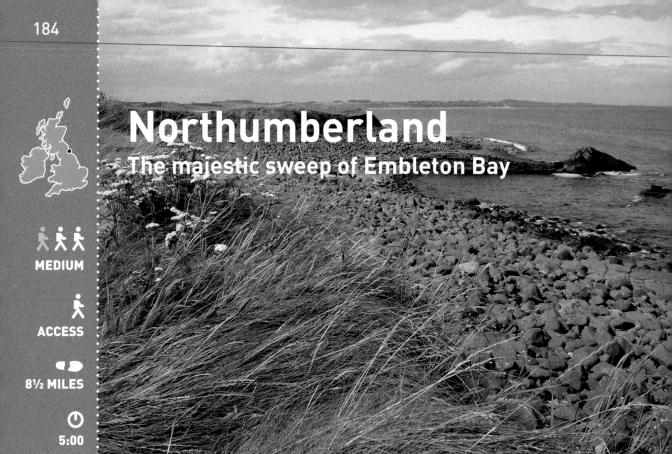

Northumberland
The majestic sweep of Embleton Bay

MEDIUM

ACCESS

8½ MILES

5:00

THE WHIN SILL

Early coalminers in the region regarded any hard rock encountered in coal seams as 'whinstone'. Anything horizontal was called a 'sill' by quarrymen. The Whin Sill is an outcrop of volcanic rock called dolerite, which was forced into the area as molten magma around 300 million years ago. As the rock cooled and shrank it formed massive natural pillars or columns with a roughly six-sided cross-section. Its escarpment outcrops in Northumberland are known as 'heughs'.

This long circular walk offers beautiful views of some of the UK's most dramatic and varied coastline. with sweeping sandy beaches, rolling dunes, high rocky cliffs and imposing castles.

1

Begin your walk from the parking area at Dunstan Steads. Here you will have sneak previews of the treats in store for later, but resist the temptation and walk away from the shore in the opposite direction. Go up the metalled road along which you have just driven and 200 yards further on take a left turn at a row of cottages. The route is marked with Public Bridleway to Dunstan Square and a North Sea Cycle Route sign to Craster.

Follow the bridleway southwards through the farm buildings and out along a concrete track. As you begin the gentle climb the majestic **Cheviot Hills** come into view on the skyline to your right, seen across the fields. As you crest the rise you will notice some Second World War concrete pillboxes and a stone-built crop store by a clump of pines. Continue along the mainly level concrete track, noticing a line of low crags on your left. With further progress these gorse-clad crags become more prominent. These are whinstone,

a very hard, resistant rock of volcanic origin and part of the geologically significant Whin Sill, found throughout the region.

② Continue along the track until you reach the farm buildings at Dunstan Square. Here you can make a short detour to view the crags of the **Whin Sill** close up and for some magnificent views out to sea. Follow the footpath sign on your left, passing eastwards through the field gate and along the stony track that edges the field. Pass through another gate and make the gentle climb into the crags. Where the track levels out you will be greeted by spectacular views out to sea, with the impressive remains of Dunstanburgh Castle standing on a rocky promontory in the distance. Savour the view here for a few moments before retracing your steps back to the farm buildings at Dunstan Square.

③ Continue along the road round the buildings and after about 200 yards go through a gate on the left, opposite a row of cottages, and onto a track bordering fields. This footpath is signposted to Dunstan. Walk south a short distance until you reach another gate. Go through this and turn left, walking along the field edge until you reach a gate in the corner of the field under some electricity cables. A short section of indistinct path takes you towards crags, and onto a wide and distinct grassy path. Turn right, pass through a National Trust waymarked gate, and walk through tall vegetation on a well-made path until a gate takes you to the roadside opposite the **Craster Tourist Information Centre**. Here there is an opportunity to buy maps and guidebooks, and get local information about forthcoming events; literature is also available about the North Sea Trail. Refreshments can be obtained here from the kiosk, including the local speciality, Craster kipper rolls. Toilets complete the amenities.

④ Return to the road and walk 500 yards into scenic **Craster Village**, where you will be greeted by a picturesque harbour, complete with lifeboat station. Turning left at the harbour, walk north past a row of cottages, following the tourist signs to Dunstanburgh Castle. Once through the kissing gate a waymark sign indicates you have joined the Northumberland Coast Path, part of the North Sea Trail. The walk continues along broad grassy paths bordering the coast, with magnificent views of Dunstanburgh Castle on its rocky promontory. The resistant dolerite rock of the Whin Sill forms the coastline here and also the crags upon which the castle is built. The sill can be examined on the shore and the roughly six-sided cooling patterns can be seen. The grassland over which you are walking forms a very distinctive feature of the region called Whinstone grassland, with a flora developed on the thin soils overlying the sill that is almost unique to Northumberland. This flora supports the caterpillars of the rare northern brown argus butterfly and if you are lucky you may spot one of them during July, after they emerge.

⑤ Continue along the broad grassy track towards the castle until you reach a kissing gate. Through the gate take the path to the left unless you are tempted to visit the castle, in which case the right-hand path should be taken. The left-hand path takes you under the crags of whinstone, following the escarpment edge until you reach the gate that opens onto **Dunstanburgh Castle golf course**. The Northumberland Coastal Path takes the left

DUNSTANBURGH CASTLE

This impressive castle was built by the ruthless Thomas, Earl of Lancaster, the nephew of King Edward II of England, to challenge the authority of the crown. Thomas was subsequently executed for treason in 1322. Later it became a Lancastrian stronghold during the Wars of the Roses, suffering significant artillery damage during the conflict.

CRASTER VILLAGE

The Craster family have lived in the area since the early 15th century, at nearby Craster Tower, and were responsible for building the current fishing harbour in 1906. A fishing port since the 17th century, Craster was also important for whinstone quarrying, supplying kerbing for London, which was transported south by sea. This industry thrived in the early part of the 20th century; nowadays the renowned kipper smokeries are the main commercial attraction.

track and skirts inland of the golf course, but take the right-hand branch onto the dunes bordering Embleton Bay. Before you reach the dunes take a look at the water's edge to see the strange rock formation known as Greymare Rock. This was formed when the limestone rocks, then buried deep in the Earth, were subjected to intense pressure and buckled during major earth movements some 300 million years ago.

6 The path now takes you onto the dunes and gives magnificent views of the golden sands of **Embleton Bay**. As you climb the dunes the vista unfolds before you. Marram and lyme grasses are important constituents, helping to stabilise the sand and minimise erosion of this fragile ecosystem. Once bound by these early colonisers other species can develop. As you walk the dunes path you could be greeted by carpets of summer-flowering bloody cranesbill or burnet rose, so characteristic of the Northumberland coastal dunes.

Continue your journey, feasting on the scenic beauty and tranquillity of Embleton Bay, a landscape that typifies the Northumberland coast. Inland from the coastline the dunes have stabilised and are extensively vegetated, forming dune heathland. You can see this to the left of the path, while on the coastal side the dunes show more bare sand with only intermittent grasses. As you now rise over the dunes more small paths open up and in late summer bracken may make them indistinct. However, your line of travel is clear. Two significant dunes give superb views of the bay with the hamlet of Low Newton visible in the distance. Approaching the holiday chalets along the waymarked track look seaward to view the limestone reefs of Emblestone: Out Carr and, closer to you, Jenny Bells Carr. Along much of the coastline of Northumberland the limestone reefs form impressive cliff features, complementing the broad sandy bays.

7 Follow the path through the dunes until you reach a gate at the **Newton Pool Nature Reserve**. You can get glimpses of the pools from the gate, home to many migrant and native bird species. Don't go through the gate, however, but turn left and follow the edge of the golf course, going through a gateway into an open field. There is an indistinct waymarker on a gatepost and the path continues alongside the field edge to another gateway, with a more distinct waymarker. Follow the path through the middle of the field and then along the field edge, climbing the small rise, with Cheviot coming into view in the distance across the fields. At the crest take a small detour right and climb to the top of Kelsoe Hill for another view of the castle and whinstone crags across Embleton Bay, with farmland in the foreground. Savour the tranquillity of this spot for a while, noting where you have travelled on this walk so far.

8 Retrace your steps back to the path, proceed down to the stone stile then left along the field edge. At a corner of the field with waymarks, ignore the left track, continuing straight ahead until you reach another stone stile. Turn left along the side road into Embleton and left again at the Dunstanburgh Castle Hotel. Follow this road up the rise and continue down towards the sea along the metalled road, past the golf course clubhouse. Now follow the footpath seawards to the footbridge over the Embleton Burn. Cross the bridge and take the path on the right, skirting the side of the golf course and going up into the **dunes**. As you walk take one last look at the castle and its crags in front of you before you descend into the gulley. Take the right-hand footpath through the golf course, back to the car park, and the end of the walk.

Northumberland

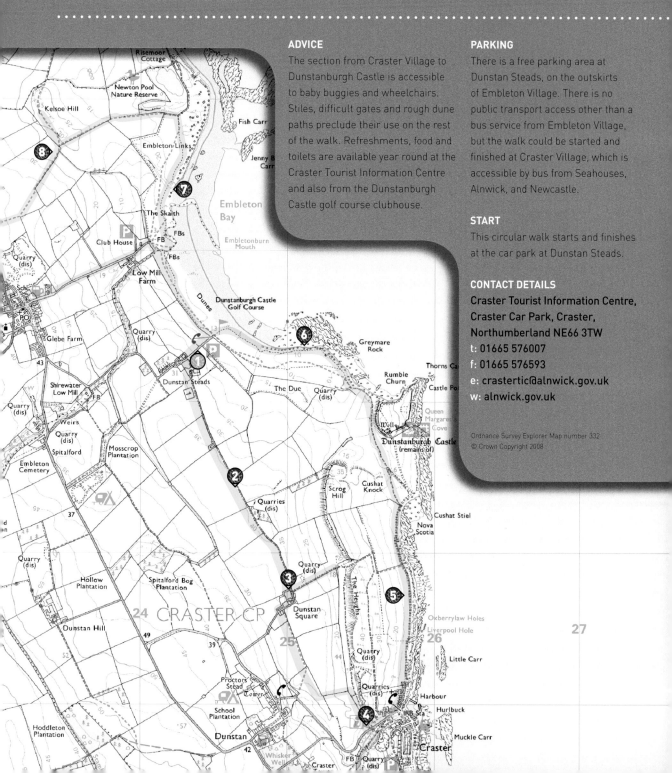

ADVICE

The section from Craster Village to Dunstanburgh Castle is accessible to baby buggies and wheelchairs. Stiles, difficult gates and rough dune paths preclude their use on the rest of the walk. Refreshments, food and toilets are available year round at the Craster Tourist Information Centre and also from the Dunstanburgh Castle golf course clubhouse.

PARKING

There is a free parking area at Dunstan Steads, on the outskirts of Embleton Village. There is no public transport access other than a bus service from Embleton Village, but the walk could be started and finished at Craster Village, which is accessible by bus from Seahouses, Alnwick, and Newcastle.

START

This circular walk starts and finishes at the car park at Dunstan Steads.

CONTACT DETAILS

Craster Tourist Information Centre, Craster Car Park, Craster, Northumberland NE66 3TW
t: 01665 576007
f: 01665 576593
e: crastertic@alnwick.gov.uk
w: alnwick.gov.uk

Ordnance Survey Explorer Map number 332
© Crown Copyright 2008

Durham
Heritage Coastal Path

MEDIUM

ACCESS

8 MILES

4:30

DENES AND GILLS

These very steep-sided valleys were carved through the bedrock by glacial meltwaters. Many began as sub-glacial drainage channels formed under the ice sheets and were then enlarged and deepened when the ice melted at the end of the last glacial period, around 10,000 years ago. They are very characteristic of the coast of north-east England.

This spectacular walk takes you through a unique environment, a rich natural and cultural asset in an area devastated by industry and then by economic deprivation as that industry declined.

. . . .

1 Begin your walk at the parking area at Limekiln Gill. Walk along the path to the sea, noting the information board and the stone tablet in the grass detailing the **Turning the Tide project**. As the mouth of the gill opens out near the large concrete blocks, a thin soil path climbs the left bank of the cliffs. This gives access to the cliff-top path and after the 20-foot climb you are rewarded with impressive views. As you walk along the cliff-top take a look to the left, noticing the grassland meadow. This is magnesian limestone grassland, which in summer is a blaze of colour, with numerous species of wild flowers growing in this rare habitat. Soon the path makes one of its many descents into the numerous denes and gills that dissect the cliff line along the coast.

2 By the cliff-top National Trust sign go right and descend the steps to the floor of **Whitesides Gill**. Turn left and walk along the

TURNING THE TIDE
project to restore the Durham Coast
is a Partnership between

A MILLENNIUM PROJECT
SUPPORTED BY FUNDS
FROM THE NATIONAL LOTTERY

Durham County Council
District of Easington
The National Trust
One NorthEast
The European Union
The Countryside Agency
Northumbrian Water

English Nature
Durham Wildlife Trust
Groundwork East Durham
The Environment Agency
Northern Arts and
Seaham Har...

path, noting the steep-sided walls of honey-coloured rock. This is magnesian limestone and represents a unique geological feature of east County Durham. Walk up the gill along the flagstone path, following the drainage culvert, and climb 50 feet to the top where the path crosses the head of the gill. Continue along the cliff-edge path, noting abundant magnesian limestone grassland on the left. On this plateau the seats make an ideal point to stop for a while and savour the views of this magnificent coastline before continuing your walk. Follow the flagged path across the head of another un-named dene and continue along the cliff-top. By now the views of cliff scenery are becoming very impressive and Beacon Hill, the turning point for the walk, is noticeable on the skyline.

❸ Soon the coastal path reaches the edge of another dene where you need to thread right along indistinct gravel tracks until you reach the edge of the cliff where steps lead you down to the beach. This is **Warren House Gill**. This valley and its immediate area boast recognition for their geology and is a Site of Special Scientific Interest (SSSI). Notice that the cliffs bordering the sides of this gorge are boulder clays, formed at the end of the last ice age. During the last glaciation, when ice-sheets moved over the area around 200,000 years ago, ice from Scandinavia carried with it pebbles of the local rocks. When these ice sheets began to melt at the end of the ice age these pebbles were deposited within the boulder clay and can be found on the beach today. Follow the gravel path into the gill, noting the vegetation. In summer parts of the gill

floor are covered by forests of huge horsetails, plants related to those that once colonised the land over 300 million years ago during the Carboniferous period.

Walk through this prehistoric landscape until you reach a stile, with a path and steps leading upwards to the cliff-top on the right. The flagged path meets another at the top and you should take the right-hand branch, which takes you back along the Durham Coastal Path. There are sand dunes here, evidence of earlier sea level changes after the last ice age. Passing two gateposts, you rejoin the limestone plateau and here, where the boulder clay capping is thin, magnesian limestone flora is abundant. This grassland habitat is also home to the Durham argus butterfly. If you are lucky you may spot these in July and August.

❹ Follow the gravel track of the coastal path along the field-edge until you reach the gorge of **Foxholes Dene**. It is deep and heavily wooded and, by looking seawards to the dene mouth, the honey-coloured side walls indicate that this has been cut into the magnesian limestone bedrock. There is no way down at this point and the path continues along the side of the dene, inland towards the railway. In places the path becomes quite overgrown and undulating, but eventually emerges alongside the railway bridge at the head of the dene. Turn right along the tarmac path, where a National Trust sign announces you are in Foxholes Dene. Follow the path along the cliff-top, ensuring that you take in the scenery towards Shippersea Point as you head north.

❺ By following the tarmac path, continue round the edge of the broad Busiers Holes, taking the right-hand gravel path by a seat to bring you back to the cliff-top, away from the railway bridge. You will notice a black oblong feature on the left across the fields. This is the site of **Easington Pit**. The cage is now all that remains of the former colliery, with the village of Easington now immortalised as the setting for the film *Billy Elliot*. Continuing northwards along the coastal path the scenery becomes

MAGNESIAN LIMESTONE GRASSLAND

County Durham has the country's only major surface outcrops of this formation. The limestone beds were laid down around 280 million years ago in a shallow, very saline tropical sea during the Permian period. The unique character of the bedrock results in alkaline soils above it that support a very special flora. This is almost the only place in Britain where this rare habitat exists, and this walk offers the best opportunity to see it. This habitat supports many rare and endangered plants. Greater knapweed, slender St John's wort, rockrose and kidney vetch are characteristic, with national scarcities such as dark-red helleborine and bird's-eye primrose represented, along with perennial flax.

⑦

⑤

ANCIENT WOODLANDS

Within the denes are preserved remnants of ancient woodland that initially developed at the end of the last ice age. While human civilisation has removed this woodland from the plateau regions, the steep topography of the denes has hampered human intervention. The denes are now a window into the past, allowing us to see the vegetation cover that clothed the landscape 7000 years ago. Ash, wych elm, hazel and hawthorn characterise this species-rich flora.

HAWTHORN DENE

An SSSI designation recognises this area's biological significance for the plants and animals it supports and the ancient woodland it preserves. The railway and viaduct were constructed around 1905, linking Seaham to Hart Junction. The Pemberton family, local 19th-century landowners, had their own private station platform not far from here.

even more majestic; the rocks and sea stacks of Shippersea Point are particularly spectacular. At a field-edge the path cuts inland towards the railway again, but it is worth making the detour along the right-hand path, which takes you close to the cliff-edge and the Point.

6 Now either retrace your steps back to the Durham Coastal Path or follow the cliff-edge path alongside the field until the two join up again at the National Trust sign for **Shippersea Bay**. On your left is the railway crossing and the bulk of Beacon Hill. Grey-coloured magnesian limestone can be seen low down in the cutting. Beacon Hill represents an ancient barrier reef, which grew in the shallow tropical waters of the magnesian limestone sea around 280 million years ago. Passing through the gate the path now continues on a species-rich grassland.

7 The gravel path alongside the railway now begins to descend and the creamy-coloured cliffs of Hawthorn Hive come into view, across the mouth of **Hawthorn Dene**. The term 'hive' is a corruption of 'hythe', meaning a safe, sheltered landing for boats. Hawthorn Dene is very impressive – a deep, wide, steep-sided gorge cut into the rock, boasting a magnificent viaduct. This is the turning point for the walk, but take the

opportunity to go into the gorge to explore it at close quarters. Follow the path under the viaduct till you see an indistinct waymarker by a track branching right. Follow this right-hand path as it winds down the side of the dene to the footbridge, then scramble down to the valley floor and turn seawards until you reach the mouth of the dene. This is an exhilarating detour with an other-worldly character to it. Once on the beach you can marvel at the natural architecture of the dene, with its steep, enclosing walls.

8 Retrace your steps to the waymarker and turn right onto the main path, which takes you into the ancient woodland on the edge of the dene. Follow this path through the woods and out onto the open grassland of **Beacon Hill**. Go through the kissing gate and walk to the top. The views from the summit are superb. Looking northwards the city of Sunderland can be seen on the skyline, and the sea cliffs and bays towards Nose Point. Southwards the cliffs and coves, stacks and shoreline down to Castle Eden Dene are spread out in magnificent splendour, with the port of Middlesbrough in the distance. Having feasted on the view, continue southwards down the hill, alongside the fence. You soon reach the corner of the field, where a strategically placed seat allows you to rest. By the seat is a kissing gate. Walking through this allows access to the stile on the left and the path seaward to the pedestrian crossing over the railway line. Descend the sloping path alongside the fence and gorse bushes. Cross the railway line with care and rejoin the Coastal Path south to make your way back to the starting point.

⑥

⑧

Durham

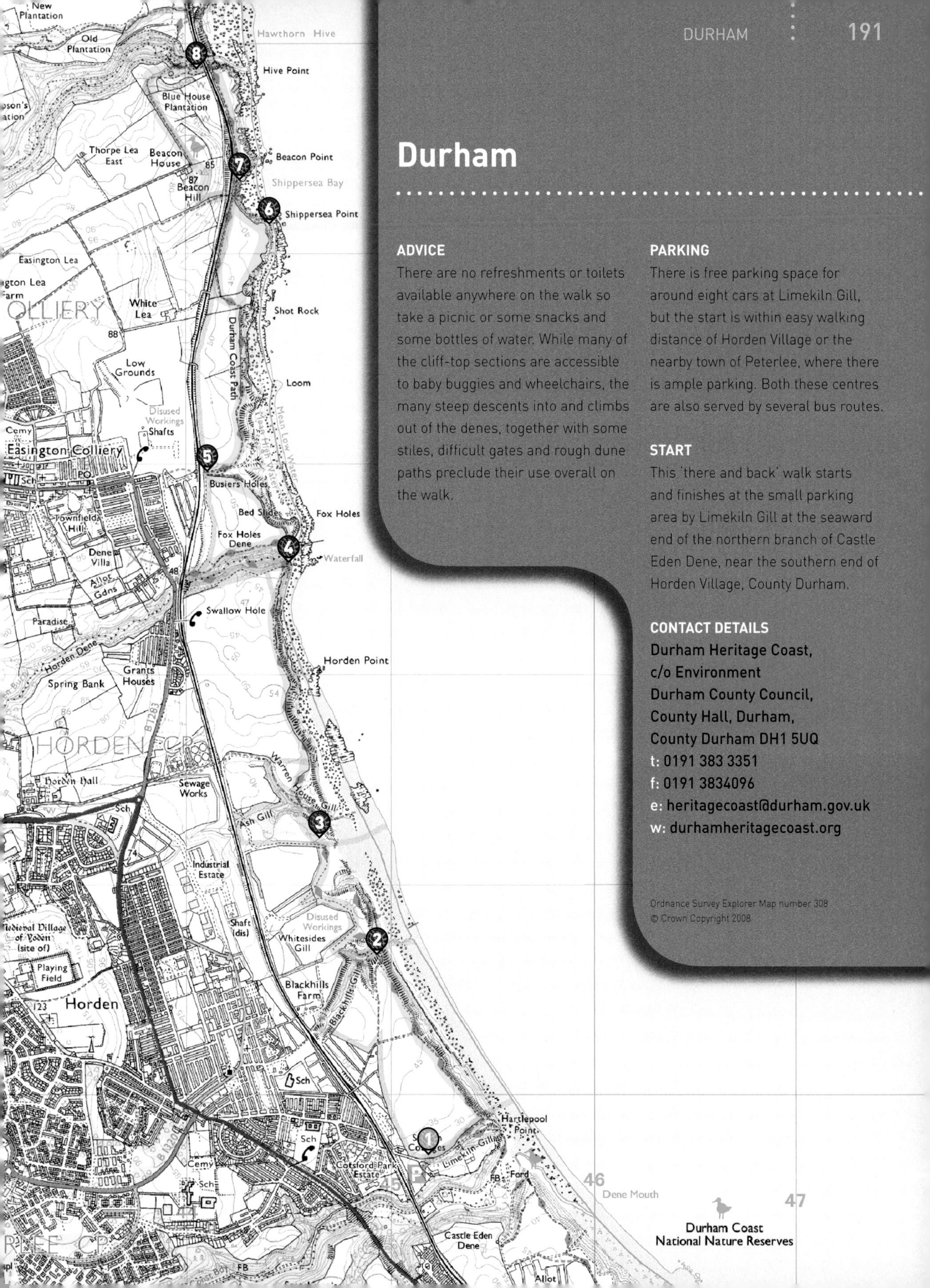

ADVICE

There are no refreshments or toilets available anywhere on the walk so take a picnic or some snacks and some bottles of water. While many of the cliff-top sections are accessible to baby buggies and wheelchairs, the many steep descents into and climbs out of the denes, together with some stiles, difficult gates and rough dune paths preclude their use overall on the walk.

PARKING

There is free parking space for around eight cars at Limekiln Gill, but the start is within easy walking distance of Horden Village or the nearby town of Peterlee, where there is ample parking. Both these centres are also served by several bus routes.

START

This 'there and back' walk starts and finishes at the small parking area by Limekiln Gill at the seaward end of the northern branch of Castle Eden Dene, near the southern end of Horden Village, County Durham.

CONTACT DETAILS

Durham Heritage Coast,
c/o Environment
Durham County Council,
County Hall, Durham,
County Durham DH1 5UQ
t: 0191 383 3351
f: 0191 3834096
e: heritagecoast@durham.gov.uk
w: durhamheritagecoast.org

Ordnance Survey Explorer Map number 308
© Crown Copyright 2008

Whitby
Living off the 'silver darlings'

EASY

ACCESS

2 MILES

2:00

This historic fishing port of Whitby in north Yorkshire has a fascinating past, which features shipbuilding, geology... and the arrival in England of Count Dracula.

1 The walk starts outside Whitby's tourist information centre. Don't be fooled by the peaceful setting and quaint skyline, Whitby's tranquil façade belies the town's prosperous history as one of the country's most productive centres of shipbuilding.

From the mid 18th to the mid 19th century, the area from the bridge to what is now the Marina car park was a bustling cluster of **shipyards**, roperies and sail yards, accompanied by the sound of the caulkers' hammers working to ensure that Whitby's ships remained watertight. Those employed in building Whitby's ships, and their employers,

lived close to the yards in dwellings that have since disappeared, but they left their imprint on the Old Town – a maze of steps, yards and houses with a distinctively nautical feel.

2 Cross the bridge and turn right into Grape Lane. The name has nothing to do with vines, it's more to do with the 'groping' that used to go on here as a result of the lack of street lighting in the 18th century.

At the far end of Grape Lane is the **Captain Cook Museum**. This imposing house, built in 1685, once belonged to John Walker, a Quaker shipping merchant. It was while lodging here that the 18-year-old James Cook

learnt his trade. During the winter months, when Walker's ships were not carrying coal from Newcastle to the Thames, James Cook lodged in the family's attic.

Cook was eager for knowledge and after several voyages 'before the mast', his studious habits were recognised. During the war with the French in 1755, he enlisted as an able seaman on the *Eagle*. After no more than a month, he was promoted to master's mate. Four years later he made master and embarked on his famed voyages of discovery aboard the *Endeavour*.

3 Turn left up Church Street to the market square. Can you imagine being able to make money from selling your own urine? Stale urine was essential to the crude processing of alum. Used as a mordant to make colours 'fast', alum was a rare substance. It was discovered and subsequently mined at Whitby from 1600.

The alkaline properties of stale urine made it an essential ingredient for alum miners. Huge vats of it were stored here and you can imagine the smell... It is said that this urine, taken from public collection points, gave birth to the saying, 'take the piss'.

But the stench of stale urine was nothing compared to the aroma produced by the bubbling tubs of whale blubber, which were found dotted around the town during its spell as a whaling port. Between 1753 and 1837 Whitby was home to some 20 whalers who earned a reputation for their technique.

However, the most unlikely reincarnation of Whitby's economy was the jet industry. Whitby jet is the fossilised remains of the monkey puzzle tree and has been used by carvers since the Bronze Age. The mechanisation of production in the first decade of the 19th century helped transform it from a minor industry into a larger employer than the town's shipyards. By 1873 Whitby had more than 200 **jet manufacturing shops**. Today the evidence of Whitby's jet industry has all but disappeared, though the Whitby Jet Heritage Centre at the end of Church Street displays the last remaining example of an authentic Victorian jet workshop.

4 Make your way back over the bridge and turn right to continue the walk. Quaffing a pint of the best Yorkshire ale in Whitby in the late 18th century along Haggersgate or **Pier Road** was a dangerous pleasure. Fine though the ale might have been, warm though the log fire was, unwelcome visitors were likely to burst through the doors of the hostelry. It would have been the press gang.

In those days conditions at sea were unspeakable and recruitment for the Navy was at an all-time low. Desperate for new recruits, it employed the strong-arm tactics of the press gang. These gangs of licensed thugs were supposedly limited to seizing seamen between the ages of 18 and 55, but they often ignored regulations.

The only hope of escape was to run. Look around you as you walk along and imagine the scene as men ran through any one of the town's 120 interconnecting yards in a desperate attempt to escape a life of misery. It was not until the end of the Napoleonic Wars that quaffing a pint in one of Whitby's many pubs became a pleasure again.

SUCCESS STORY

The town's success as a centre for shipbuilding was based on its reputation for building vessels that were both durable and strong. Among many ships built here were Captain James Cook's three famous 'cats', the *Endeavour* (1764), *Resolution* (1769) and *Adventure* (1770). These flat-bottomed boats, originally designed for the alum industry, were perfect for Cook who often needed to 'take the ground' safely in unknown waters. In the 19th century, the city's most famous whaling captain, William Scoresby, and his son, William Junior, undertook their first Arctic adventure in a Whitby-built vessel. They succeeded in penetrating the Arctic ice further than anyone else.

SMOKING

Before the coming of the railway, which enabled the fresh fish to be taken away as soon as it was landed, alternative ways of preserving this precious product were needed. A number of notable fish smokers were established in the town. Perched at the foot of the cliffs beneath St Mary's Church, Fortunes, in Henrietta Street, is one such business. The puffs of smoke that turn to clouds as the fires belch out in the smokehouse, have been going for over 120 years. Visit them – the smell is heavenly.

DRACULA ARRIVES IN WHITBY

But the strangest of all, the very instant the shore was touched, an immense dog sprang up on the deck from below, as if shot up by the concussion, and running forward, jumped from the bow to the sand.

Bram Stoker used a description of the real-life wrecking of the *Dimitre* to 'transport' his character Count Dracula to England in the guise of a dog.

5 Follow the sound of gulls, and the unmistakable smells of fish and brine, and you will find yourself at the town's quayside. Fishing has been in the blood of the people of Whitby for generations and herring, the 'silver darlings' as they were known, has always been the fish of choice. Traditionally, August and September saw the **herring boats** gather in the harbour in such great numbers that, at times, you could walk from one side of the harbour to the other across their decks. And the women-folk played their part too. Fishwives, as they were affectionately called, were paid 1s 6d (7½p) a day for their onerous and odorous work, counting the fish on board the boats.

Of course, the buyers came off best. Herring were sold by the 'last', nominally, 10,000 fish. But with four herring making a 'warp', 33 warps making a 'hundred' and 100 hundreds making a last, it meant that the fisherman was paid for 10,000 herring while the buyer received 13,200 fish.

6 From the fish quay, head past the Victorian bandstand and onto West Pier. Look up at the ruined **Gothic abbey** on the cliffs above the town. It was high up on these cliffs that Bram Stoker, the Irish author, penned much of his novel *Dracula*. Today the haunting ruins of the seventh-century abbey still dominate the landscape for many miles.

And there are those, even today, that relive the Gothic life to the full. Twice a year, in April and at Halloween, Whitby becomes 'Goth City', with music, performances and celebrations of gothic life and times.

7 Now go back along the pier and take the first right, up the winding Khyber Pass. Smuggling was rife in the 18th century and the port of Whitby, with its growing maritime traffic, was ideal as a receptacle for all manner of contraband.

One story puts Skinner Street at the hub of this clandestine trade. Captain Harold Hutchinson was ordered to Whitby in the late 1700s to act against the smugglers. He was quickly made customs officer and began to avail himself of certain items of contraband. Such was the extent of his dealings, he amassed considerable wealth and was able to build a fine dwelling in Skinner Street, which became known as 'Harold's Mansion'.

Next to you at this point is a **whalebone arch** – a reminder of the men who once hunted the whale. Two hundred years ago the women of Whitby would climb up and stand on this bleak headland to watch their husbands and sons leave for the icy north.

8 From the arch, turn round and walk in the opposite direction, until you reach **Royal Crescent**. The arrival of the railway is an integral part of Whitby's history, heralding a new source of income for the town through tourism. And it was largely down to the vision of one man – George Hudson – who

built Royal Crescent on the considerable proceeds of his business.

An entrepreneur from the East Riding of Yorkshire, Hudson invested in the newly established North Midland Railway. He proposed a rail link between York and the West Riding, which took in the seaside delights of Whitby. The link, which opened in 1839, opened up the town as a favourite destination for large numbers of factory workers from the West Riding and beyond. Day trips became possible and affordable.

Guesthouses and hotels were developed on the west cliff for those staying longer.

Though the crescent is only half built as its full development was cut short when Hudson ran out of money, his legacy of tourism lives on and has been essential to the town's fortunes ever since.

From Royal Crescent, head back towards the whalebone arch and take the first right. Keep walking straight down, towards the church, and turn left when you reach the end of Skinner Street. Walk past Silver Street and take the next right – Brunswick Street. Walk down Brunswick Street until you reach three churches all on one corner. Bear left here and head back towards your starting point.

FRANK MEADOW SUTCLIFFE

Whitby has always attracted artistic tourists. The fishing industry and the harbour environment were their key subjects. The photographic work of Frank Meadow Sutcliffe is perhaps the most celebrated and much of his work can be seen in the Sutcliffe Gallery in Flowergate.

Whitby

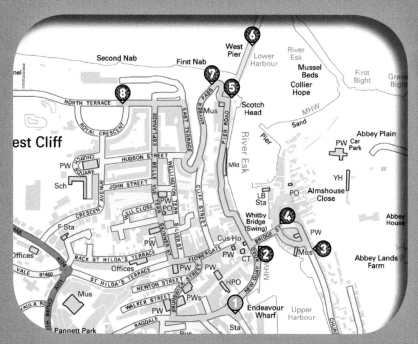

Ordnance Survey Explorer Map number OL27
© Crown Copyright 2008

ADVICE

The whole walk is on pavement and tarmac. However, cobbles feature in one part of the walk and the Khyber Pass is steep but still accessible for wheelchairs and buggies.

PARKING

There is a short-stay, pay-and-display car park in Cliff Street close to the start of the walk or a long-stay car park over the bridge in Church Street.

START

The walk begins outside the Tourist Information Centre.

CONTACT DETAILS

Whitby Tourist Information Centre, Langborne Road, Whitby, North Yorkshire YO21 1YN
t: 01723 383636
f: 01723 383604
w: whitbyonline.co.uk

Flamborough Head
Seabirds on display

MEDIUM

ACCESS

4½ MILES

4:00

CHALK

Flamborough Head and Bempton Cliffs are composed of chalk that was deposited between 65 and 80 million years ago during the Cretaceous period. It is formed from the microscopic skeletons of tiny marine algae that lived at a time when the climate was considerably warmer and sea-level was between 200 and 300 feet higher than it is today.

This walk takes in the cliffs of Flamborough Head and Bempton, which stand up to 400 feet high and are home to some of the largest colonies of nesting seabirds in England.

1 Flamborough Head is one of the most prominent features of the East Coast. It juts out into the North Sea and was defined as part of the Heritage Coast in 1979. In the Iron Age it was effectively isolated by the construction of Dane's Dyke. As well as the distinctive white cliffs there are coves, sea caves and stacks along the coastline. At the top of the cliffs above the chalk is a thick layer of brown clay (known as 'till'), which was deposited by glaciers during the last ice age some 18,000 years ago. As the cliffs below are worn away by the waves, the clay often falls into the sea as huge landslips

Start the walk at the Flamborough Head car park near the modern **lighthouse**. The older Chalk Tower light beacon can be seen in the distance. Follow the signs to the cliff-top path, descending some steep steps towards Selwick Bay that can be accessed, although the steps down to the bay itself are somewhat dilapidated. Admire the magnificent view of the high chalk cliffs with their almost horizontal layering and patches of grey flints.

2 Continue along the cliff-top path along the northern side of the golf course. After about a mile you will start to pass a number of interesting examples of **sea stacks**,

the result of constant erosion by the waves of the North Sea. After another half mile you will reach the cliffs overlooking North Landing beach. From here you can observe a row of sea caves exposed at the waterline on the far side of the bay. At the top of the boat slipway are a couple of cafes, toilets and a car park. It is also possible to take a half-hour boat trip to Smuggler's Cave and the bird sanctuary or a three-hour fishing trip (daily from 1.00 p.m. – rods and bait provided).

At this point, if you do not wish to proceed to Bempton Cliffs, it is possible to return on a circular route to Flamborough Head along the road through Flamborough village.

3 Descend the steep path down onto the beach at **North Landing**. Here you can explore the caves and rock pools exposed at low tide or simply enjoy a picnic on the sand. This is a good place to investigate the chalk close up and to see the layers and nodules of grey flint. Flint is composed of silica, which was probably derived from the spines of Cretaceous sponges. This dissolved and then was resolidified in areas where there was decaying organic matter (e.g. in burrows and around dead marine organisms such as sea-urchins). Look out for birds' nests in the cliffs. At the back of the beach near the cliffs are piles of glacial clay that have slipped down from above the chalk. These contain boulders and pebbles of material deposited by melting glaciers at the end of the last ice age.

4 Return to the cliff-top path and proceed past **Thornwick Bay** towards Bempton Cliffs. There is a car park, toilets and small cafe at the far end of Thornwick Bay. From here it is also possible to return on a circular route to Flamborough Head by returning to the road through Flamborough village.

5 From Thornwick Bay to Bempton Cliffs is a fairly stiff walk, but with some stunning cliff-top scenery and plenty of opportunities for birdwatching and photography. The cliffs host numerous **nesting colonies** of many different types of seabird, including gannets, guillemots, kittiwakes, puffins, razor bills, fulmars, herring gulls and shags. The best time to observe the colonies is in late spring and summer when the breeding season is in full swing. By the autumn almost all the seabirds have departed and breeding is finished, except for the gannets. However, the autumn migration can be well observed with the arrival of migrants such as short-eared owls, warblers, whitethroats, stonechats, wheatears and redstarts. Scarce

species such as red-backed shrikes and barred and icterine warblers may also be spotted. Offshore, Arctic and great skuas, and Manx and sooty shearwaters, can be seen in ideal weather conditions (strong north-westerly winds). The winter is the least busy time of the year. The cliffs attract very few birds, except for occasional herring gulls and fulmars, though gannets usually return in January.

As well as birds, a range of common butterflies can be seen on sunny days, along with day-flying moths such as burnet moths,

LIGHTHOUSES

The modern lighthouse was built in 1806. It is 85 feet tall and was built without using scaffolding. An older beacon light tower (the Chalk Tower), dating from the 1660s, stands further back and is the only known example in England.

CAVES AND SEA STACKS

Caves are formed as the sea erodes weak cracks or joints in the cliffs. If the cliff juts out, a cave may be worn away into a rock arch, which, on collapse, leaves a pillar of rock cut off from the rest of the cliff. This is known as a stack. This process of erosion can take hundreds of years.

THE PUFFIN

This comical bird can easily be identified by its black back, white underparts, large white cheeks, red and black eye markings, orange legs and brightly coloured bill. It nests in burrows or under boulders. Both parents usually incubate a solitary egg and the departure of all adults usually takes place within a few days.

cinnabars and occasionally hummingbird hawk-moths. Cliff-top flora is dominated by red campion, black knapweed, various thistles and orchids, including common spotted, northern marsh and pyramidal.

6 As you walk towards Bempton Cliffs you will pass a small wooden **carving of a puffin**. Looking inland from here you will be able to see the wooded ridge of Dane's Dyke stretching southwards across the fields. Dane's Dyke is a massive ditch and bank earthwork running some 3 miles across the Flamborough headland, from Cat Nab in the north to Cliff Plantation in the south. It forms a formidable barrier, which effectively cut off the headland and must have contained all the resources to support a sizeable population. Despite its name, Dane's Dyke was not constructed by the Vikings; opinions on its age are divided, from Neolithic to Bronze or Iron Age. The only recorded excavation took place in 1879 when over 800 prehistoric flint artefacts were found, together with a few fragments of pottery. Some have even noted similarities to other post-Roman earthworks.

7 Because of its national importance Dane's Dyke is a protected monument. Much of it is privately owned, though it can be easily viewed from the cliff-top path and public roads. The only part open to the public is to the south at the Dane's Dyke Countryside Site off the B1255 road from Sowerby to Flamborough Head. Here a long, well-preserved length of Dane's Dyke is visible and a number of nature trails are accessible on foot.

8 Passing Dykes End at Cat Nab, continue walking along Bempton Cliffs, passing several useful viewpoints for birdwatching, until you reach the Bempton Visitor Centre. This marks the end of the walk, but the visitor centre is well worth a visit for gifts and refreshments. Guided walks are available, as well as binocular hire and advice on optical equipment for birdwatching and photography. There is also a bird-feeding station and picnic area. In the

summer, around the car park area, you may be able to spot migrant hawker and common darter dragonflies.

The best way to return to the starting point is to go back the way you came. It is possible to complete part of the walk as far as North Landing or Thornwick Bay and then return along an inland footpaths or roads, though there is little of interest to see here. See Advice for details of a bus service if you are too tired to walk.

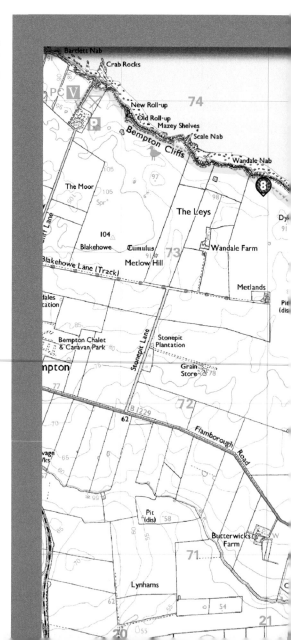

Flamborough Head

ADVICE

This is a linear walk so please be prepared to walk back unless you have made other arrangements. The walk is most suitable for adults and older children, since it is rather too far for small children. The walk is not suitable for wheelchairs or baby buggies, though there are several car parks along the way from which access to some of the viewpoints is possible. Access onto the beach areas is difficult due to steps or steep slopes. There is some wheelchair access to the beach at North Landing, but this is still very steep. Although the cliffs are well fenced the narrow cliff-top path can sometimes be very slippery and windswept so close supervision of younger children is advised. The cliffs are also liable to crumbling and collapse. The route is accessible all year round, although the best time to observe the bird colonies (particularly puffins) is between April and August. Well-controlled dogs are allowed on the beach but must be kept on a lead along the cliff-top path.

PARKING

There are pay-and-display car parks at either end of the walk. There is a public bus service (No.501) that operates between Flamborough Head and Bempton Nature Reserve on Sundays and Bank Holidays in July and August, but it is advisable to check the timetable (on the noticeboard in the car park near the lighthouse at Flamborough Head) before embarking upon the walk.

START

The walk starts at Flamborough Head and ends at the Bempton Nature Reserve where there is a useful shop and visitor centre.

CONTACT DETAILS

Bridlington Tourist Information, 25 Prince Street, Bridlington, Yorkshire YO15 2NP
t: **01262 673474**
e: **bridlington.tic@eastriding.gov.uk**
w: **iknow-yorkshire.co.uk**

Ordnance Survey Explorer Map number 301
© Crown Copyright 2008

Hull
Walking on water

MEDIUM

ACCESS

2½ MILES

2:00

This walk will help explain why few places in the country can claim such a rich and impressive maritime history as Hull.

The walk starts outside the BBC Open Centre in Queen's Court. Walk away from the building into **Queen's Gardens**. For a time this was the site of the largest dock in Britain. Built in 1775, it cost under £65,000 and was officially opened on 22 September 1778.

It was initially used to export manufactured goods from Yorkshire, the Midlands and Lancashire, and import raw materials from Europe. As trade increased merchants moved from the High Street vicinity to settle in this part of town.

After the discovery of Greenland by Sir Hugh Willoughby, the first whaling ships appeared in Hull in 1598. The trade reached its peak at the beginning of the 19th century when the dock had more than 60 whalers and was the largest fleet in Britain. In 1820 these vessels caught 688 whales, producing 8000 tonnes of whalebone.

Whaling was a dangerous occupation. Between 1818 and 1869 no fewer than 800 ships from Hull were lost at sea. Whaling declined towards the end of the 19th century, and the last vessel, the *Diana*, left Hull in 1869 and never returned.

By the 20th century more docks had opened in the city and shipping had moved away. Queen's dock finally closed in 1930 and was filled in after 150 years of activity. Look carefully and you can still make out the

shape of the dock in the walls and buildings around you.

2 Walk to end of the park and turn right onto Wilberforce Drive. Cross the two pedestrian crossings and you'll be at the top of Lowgate in front of the Magistrates Court. Turn left into Chapel Lane. At the end turn left again. **Wilberforce House** is on your right.

William Wilberforce, born in this house on 24 August 1759, was the son of a wealthy merchant. He studied at Cambridge University and in 1780 became MP for Hull. His Christian faith prompted him to become interested in social reform, particularly the improvement of factory conditions in Britain. He also lobbied for the abolition of the slave trade and for 18 years regularly introduced anti-slavery motions in parliament. In 1807 the slave trade was finally abolished in the British Empire, but this did not free those who were already slaves.

Wilberforce retired from politics in 1825 and died on 29 July 1833, shortly after the Act to free all slaves passed through the House of Commons. In 1906 the house was opened as a museum in celebration of his humanitarian achievement.

3 Retrace your steps and carry on walking down the **High Street**. From medieval times until the 19th century the town's merchants built houses along this narrow street and it was from here that the business of the port was conducted.

The High Street also had a number of public houses, brothels and inns. Ye Olde Black Boy is one of the few remaining old pubs in Hull. It initially opened as a pipe shop in 1720 and has served a number of different purposes throughout the years, including a coffee shop and a brothel. It became a pub in the 1930s.

The name is thought to be a reference to a Moroccan boy who worked in the building in the 1730s. These establishments were popular places for the press gangs to operate in.

4 Continue walking along the High Street. Turn left into Scale Lane Staith. Walk to the end and turn right. You are now on the banks of the **River Hull**. The story of Kingston upon Hull begins here. Habitation of the area is thought to date back to the 11th century when this was a small settlement called Wyke.

Wyke was owned by the monks of the Cistercian Abbey of Meaux until January 1293 when King Edward I bought it to establish a military base. It was renamed King's Town or Kingston and given a Royal Charter. Even during those early days, the settlement thrived due to the exportation of wool from Yorkshire. In 1332, Edward III granted the town a new charter and the right to elect a mayor. The first mayor was William de la Pole, a wealthy Hull merchant later knighted by the king.

5 Carry on past the tidal barrier for about a mile until you get to Nelson Street, where you will see the **Victoria Pier**. Before the construction of the Humber Bridge the rushing

HULL'S FAVOURITE SON

At the far end of Queen's Gardens, on the Hull College forecourt, stands a statue of William Wilberforce. The monument was built in 1835 and is 102 feet high. It was originally placed on a site between Beverley Gate and the Ferens Art Gallery, but the statue was moved to its present location in 1935 because it was seen as a traffic hazard.

THE GHOSTS OF YE OLDE BLACK BOY

Regular drinkers at this pub are said to have photos of ghostly apparitions; one was grabbed round the neck by a pair of ghostly hands appearing from the bar wall. Rumour has it that a landlord's dog was apparently so traumatised by spending a night downstairs in the bar that it had to be put down.

FLOOD DEFENCELESS

Just downstream is where the River Hull meets the Humber estuary, with its swift currents and open waters. But even here, away from the mainstream, people suffered frequent floods. As you continue down the river you will pass under a massive modern flood barrier built in 1980 to protect the town from high tides and storm surges. However, the barrier was unable to do its job in June 2007 when unseasonal rainfall caused millions of pounds worth of damage.

THE HUMBER BRIDGE

On a clear day, looking past the Marina and Albert Docks you can see the Humber Bridge in the distance. When the bridge opened in 1981, it was the longest single span suspension bridge in the world and remained so until the construction of Denmark's Storebelt Bridge in 1998.

grey waters here were navigated every day by small boats taking passengers and produce to New Holland on the opposite bank.

The Pier opened in 1847 and has seen many changes during its lifetime, particularly in the last two decades, which have seen the development of the marina and the construction of new office blocks. The Minerva pub continues to thrive. However, there is still evidence of decline in the adjacent streets and regeneration plans for the area, including a new site for the fruit market, continue to take shape.

On the left you can see a strange-shaped building that looks like an abstract submarine. This is the Deep, an aquarium designed by Sir Terry Farrell. It took three years to construct and opened in 2002. It houses a variety of sea life, including sharks.

6 Carry on walking along the waterside. Turn right at the Minerva pub into Minerva Terrace. Continue walking along the dockside until you reach the *Spurn Lightship*.

For more than three decades this lightship lay anchored at the mouth of the treacherous Humber Estuary, its light and foghorn marking the treacherous shoals, where some 30 vessels have been wrecked. She was built in Goole in 1927 at a cost of £17,000. Having no engine, she was towed into position on the estuary on 17 November 1927.

Built from steel, the ship is divided into seven watertight compartments and constructed in such a way that she is rendered 'practically unsinkable'. She remained in service until November 1975. In 1983, the city council restored the lightship to its original condition and moored her here at the marina and she has been on display to the public ever since.

7 Cross the dual carriageway using the pedestrian crossing on the other side of Humber Dock Street. Almost in front of you is Dagger Lane. Walk down this cobbled street and then take the second right into Prince Street. At the end is **Trinity Square**.

Holy Trinity is one of the largest parish churches in England. Its tower stands 150 feet high and houses a ring of 15 bells. The church itself is 285 feet long and 96 feet wide with a total area of 25,640 square feet.

Parts of the building date back to the 13th century, but between 1841 and 1872 it underwent a major renovation. William Wilberforce was baptised here and William de la Pole's family tomb is adjacent to the chancel. In front of the church stands a statue of the famous Hull poet, Andrew Marvell.

On the right-hand side of the church is one of the oldest grammar schools in the country. It was built in 1583 to provide education for the sons of wealthy merchants. Former pupils include William Wilberforce, Dr Thomas Watson, the Bishop of Tasmania, poets Andrew Marvell and William Mason, and Dr Isaac Milnere, the Dean of Carlisle.

8 With your back to the school, turn left into Posterngate. At the end, turn right into Princes Dock Street and walk to the top. To the left you will see the excavated ruins of Beverley Gate. This was part of a series of defensive walls first built in the 14th century by Edward II. Three hundred years later the town still had a reputation for the strength of its fortifications. These were further enhanced by Charles I after the Scottish War so that he could safely store his considerable magazine of arms and ammunition.

In early 1642, as relations between Charles and Parliament deteriorated, possession of the magazine became increasingly important. Sir John Hotham, appointed Governor of Hull by Parliament, was sent to the town to defend it against the king. In April the king travelled to Hull with the intention of seizing the arms.

One rainy Saturday morning in May the king approached the town, Sir John Hotham ordered the gates to be closed and a standoff began here at Beverly Gate. Unable to gain access to the town, the king retreated, but a few months after this show of defiance the English Civil War began.

It's only a short walk from here to Queen's Gardens, where you started.

Hull

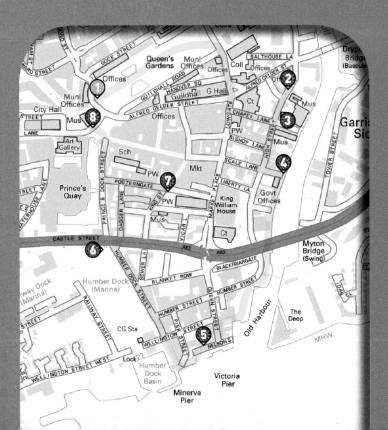

ADVICE

The majority of the walk is on pavement and cobblestones; wooden decking features on one part of the route. The whole walk is accessible for wheelchair users and parents with buggies.

PARKING

There are pay-and-display car parks in Dock Street and Lowgate, both of which give easy access to Queen's Gardens.

START

The walk starts from the BBC Open Centre on Queen's Gardens and takes you through Hull's old town and the marina.

CONTACT DETAILS

Hull Tourist Information Centre,
1 Paragon Street, Hull, Yorkshire HU1 3NA
t: 01482 223 559
e: tourist.information@hullcc.gov.uk
w: www.hullcc.gov.uk

Ordnance Survey Explorer Map number 293
© Crown Copyright 2008

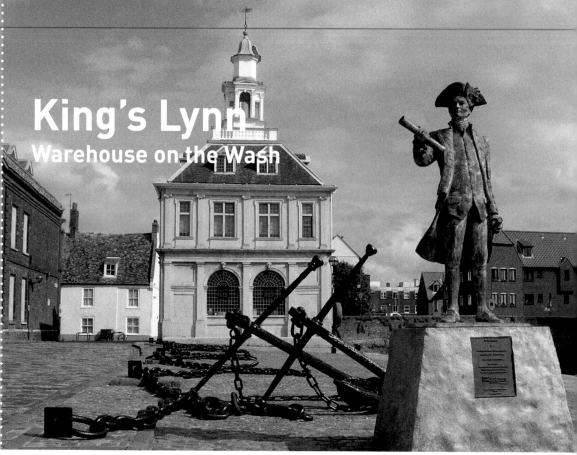

King's Lynn
Warehouse on the Wash

EASY

ACCESS

1½ MILES

2:00

Sometimes referred to as the 'Warehouse on the Wash', King's Lynn was once Britain's third busiest port and reminders of its seafaring past can be found all along this walk.

Situated on the River Great Ouse, King's Lynn is rich in heritage, with a maritime history dating back to the 12th century. The town is characterised by its cobbled lanes and medieval merchants' buildings, a legacy of its prosperous trading history.

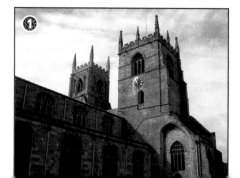

1 The walk starts at **St Margaret's Church**. Founded by the first Bishop of Norwich, Herbert de Losinga, in 1101, the church of St Margaret is now one of the largest town churches in the country. It has been substantially redeveloped over the centuries, with only the base of the south-west tower remaining from the original building. The spire of this tower was lost in 1741, when a storm sent it crashing down into the nave. A major rebuild was required, which took three years and resulted in the structure that stands today.

Walk east along the Saturday Market Place by the side of the church and continue

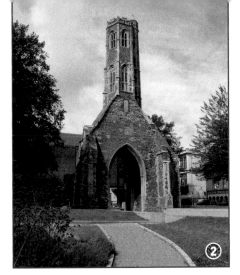

straight ahead along St James Street. After crossing Tower Place, on your right is a former cinema. This was built on the site of the Theatre Royal, which was destroyed by fire in 1936. Turn right through a gate into Tower Gardens.

2 The gardens are dominated by **Greyfriars Tower**, which is all that remains of a Franciscan monastery founded in the 13th century. After the dissolution of the monasteries during the reign of Henry VIII, most of the buildings were demolished. The tower remained as it was a useful guide for sailors returning to the port and is now one of only three such towers in England.

Walk through the gardens, which were planted to commemorate the coronation of King George V, and exit via a gate on the eastern side. Turn right and continue to the corner of London Road and Millfleet.

3 The imposing red brick building on the corner is the **public library**, which was built in 1904 and was largely funded by the philanthropist Andrew Carnegie. Walk down Millfleet and halfway along the street, on the left-hand side, you will see a red brick enclosure. This is the Millfleet Burial

Ground, where a number of Dutch Jews who lived in King's Lynn in the 18th and 19th centuries are buried. Cross Tower Place at the pedestrian crossing and the road now becomes Stonegate Street. Continue along the street to a roundabout.

4 Take Nelson Street, the road directly ahead. This bends round to the right, following the 12th century waterline. When the riverbank silted up in the 14th century, merchants built mansions on the western side of the street. Many of these had shops facing the road with warehouses to the rear, which stretched down to the river, affording easy access to their owners' ships. A good example is **Hampton Court**, which can be found on the left at the end of the street. Opposite Hampton Court, the timber-framed house on the corner is a former inn known as the Valiant Sailor. It was named to commemorate the deeds of Jack Crawford, who in 1797 nailed the colours of HMS *Venerable* to her mast during the battle of Camperdown.

Turn left into the cobbled St Margaret's Lane. The building on the right-hand side is the Hanseatic Steelyard, a warehouse which was built in the late 15th century ('steelyard' comes from the old German word for sample yard). The Hanseatic League was a guild of Baltic merchants that grew up in medieval times to protect shipping and trade throughout the region. This is the only remaining Hansa building in any of the eight English ports which were used by the merchants.

5 The lane leads you down to the River Great Ouse. Turn right and walk along the quayside with the river to your left. Look out for Clifton House's five-storey Elizabethan tower, which was used by merchants seeking the first glimpse of their incoming vessels. At the end of South Quay, cross over a small wooden footbridge to Purfleet Quay.

By the waterside is a **stone compass** that bears the names of famous sailors and explorers associated with King's Lynn. Walk along Purfleet Quay, past the statue of

RIVER GREAT OUSE

The River Great Ouse is the fourth longest river in Britain and takes its name from the Celtic word for water. It flows through the flat landscape of East Anglia and enters the Wash just north of King's Lynn. There has been a passenger ferry service crossing the Great Ouse at King's Lynn for over 700 years.

MARKETS

The first official market in King's Lynn was authorised in 1101 and is still held every Saturday next to St Margaret's Church. In 1537 the town was granted the right to hold two weekly markets. The larger Tuesday Market Place hosts the second market and is also home to the historic Lynn Mart, a two-week-long fair that opens on Valentine's Day each year.

Captain George Vancouver who was born in King's Lynn in 1757. He was a British naval officer and navigator who charted large stretches of the Pacific coastline of North America. At the end of the quay stands the Custom House. Built in 1683, from Dutch-influenced designs by Henry Bell, the Custom House is one of King's Lynn's most famous landmarks. Originally used as a merchants' exchange, it became the town's Custom House in 1718 and remained in use by Customs and Excise until as recently as 1989. This elegant building now houses the tourist information centre. Turn left and walk along King's Street until you reach the Guildhall of St George on the left-hand side of the road.

6 The Guildhall of St George is the older of the town's two guildhalls. Dating back to 1406, it is the largest and oldest hall of a merchant guild remaining in Britain. In the past the building has been used as a theatre (where Shakespeare is believed to have performed), a warehouse, a courthouse and an armoury for Royalists during the Civil War. The Guildhall and the warehouses to the rear now house the King's Lynn Centre for the Arts. Continue along King's Street until the road opens out into the Tuesday Market Place. Walk along the left side of the market place, passing the Baroque façade of the Corn Exchange.

At the far end of the market place look out for the **heart shape** above one of the first-floor windows of the red brick house. This supposedly marks the spot where, in 1616, the heart of Mary Smith, an alleged witch, hit

the house of her accuser when it burst from her body as she was executed. Continue on out of the Tuesday Market Place and walk to the end of St Nicholas Street.

Diagonally opposite is St Nicholas's Chapel. Dedicated to the patron saint of fishermen and sailors, this church is the largest 'chapel of ease' in England. It was built when the town expanded to the north in the 12th and 13th centuries to serve those parishioners of St Margaret's who resided in this new area. Turn right and walk for a short distance down Chapel Street to the timber-framed Lattice House. This 15th century building was restored in 1982 and is now a public house.

7 Turn right and walk along Market Lane, which brings you back out onto the Tuesday Market Place between Ye Olde Mayden's Heade pub and the Duke's Head Hotel. Turn left and walk past the hotel to join the pedestrianised High Street where modern stores now occupy the Georgian and Victorian buildings.

When you reach a crossroads, turn right down Purfleet Street, following the signpost to the Custom House. At the end of the street, with the Custom House opposite, turn left onto Queen's Street. Carry on, crossing the entrance to a car park, and on the right-hand side of the street is Clifton House. The front of this merchant's house is notable for the twisted pillars which frame the door. In the rear courtyard is the **lookout tower** which was visible from the South Quay earlier in the walk.

8 Further along Queen's Street, on the left, are the fortress-like Burkitt Homes, built as almshouses in 1909. Immediately opposite is Thoresby College, which was founded in the early 16th century by a prominent merchant called Thomas Thoresby. It originally housed chantry priests of the Guild of the Holy Trinity, but now contains a youth hostel, a public hall and flats for the elderly.

As St Margaret's Church returns into view and Queen's Street becomes the Saturday Market Place, on the left is the **Guildhall of the Holy Trinity** with its flint chequerboard front. The original hall with its large arched window was built in 1423 to house the Trinity Guild. The entrance porch, which bears the coats of arms of Elizabeth I and Charles II, was added two hundred years later. The hall to the left of the entrance porch was constructed in 1895, completing what is now the Town Hall complex. Next door to the Guildhall is the old Gaol House, built in 1784 as a residence for the town gaoler and containing cells to the rear. The design of the entrance doorway, with its manacles and chains, was influenced by the architecture of Newgate Prison in London. Cross the Saturday Market Place to return to the front of St Margaret's Church.

GUILDHALLS

A guild is a body of people with common interests. Medieval in origin, guilds were first formed for religious and charitable purposes. From the 14th century, trade and craft guilds controlled economic life in towns. They were self-regulating bodies that set standards and suppressed competition. Guild members met in buildings known as guildhalls.

King's Lynn

ADVICE
The route is fully accessible to anyone. There is a short cobbled section, but generally the going is smooth and flat.

Public conveniences, including disabled toilets and baby changing facilities, can be found in several locations around the town and are clearly signposted.

PARKING
Park in the pay-and-display car park in Church Street, which is located to the east of St Margaret's Church. There are also plenty of other car parks close to the route of the walk.

START
The walk starts and ends in front of St Margaret's Church.

CONTACT DETAILS
**King's Lynn Tourist Information Centre, The Custom House, Purfleet Quay, King's Lynn, Norfolk PE30 1HP
t: 01553 763044
f: 01553 819441
e: king's-lynn.tic@west-norfolk.gov.uk
w: west-norfolk.gov.uk**

Ordnance Survey Explorer Map number 250
© Crown Copyright 2008

North Norfolk Coast
An area of constant change

EASY

ACCESS

5 MILES

3:00

SAILING BARGE 'JUNO'

CLEY WINDMILL

Cley Windmill dates from the early 1700s and remained a working mill until 1919, after which it fell into disrepair. In 1921 most of the working parts were removed and it was converted into a holiday home. Following further renovation in 1983, the windmill is now used as a guesthouse. It remains one of the best known landmarks on the north Norfolk coast.

Explore the marshes and maritime past of Norfolk with this circular walk along part of the North Norfolk Coast Path, through the former trading ports of Blakeney, Cley-next-the-Sea and Wiveton.

The north Norfolk coast is an area of constant change, where the boundary between land and sea is continually shifting. This walk takes you along a small section of the North Norfolk Coast Path, through the salt and fresh marshes that have formed in the shelter of Blakeney spit, and the villages of Blakeney, Cley-next-the-Sea and Wiveton.

① Formerly a bustling trading port, Blakeney is now a popular centre for recreational sailing and the walk starts at the **harbour**. From the car park, take the raised path, signposted Norfolk Coast Path, with the fenced off wildfowl conservation area to your right. As the path continues, it divides the Morston salt marshes to the west from the fresh marshes, known as the Blakeney Freshes, that extend across to Cley.

These different habitats are both havens for wildlife, attracting a diverse range of flora and fauna. As well as the large resident bird population, spring and autumn see many migratory visitors, such as avocets, marsh harriers and spoonbills. Look out, too, for plants such as marsh samphire, yellow-flowered winter cress and common sea

lavender, which covers the salt marshes in a purple haze in summer.

2 As the path turns east, take the right fork, following the arrow on the marker post. The path leads you increasingly closer to a narrow ridge of shingle which stretches out westwards into the sea and terminates at **Blakeney Point**. As well as continuing to lengthen, this three-and-a-half mile spit is also moving landwards at a rate of around 3 feet a year. The inland migration of the spit actually threatened to block the original channel of the River Glaven, putting both the Blakeney Freshes and nearby Cley at risk of flooding. To alleviate this, in 2005 the Environment Agency diverted the river to a new channel, some 200 yards south of the original course. The coast path was also moved inland to run along the south bank of the realigned river.

3 Between the path and the shingle ridge lie the remains of a medieval building. Although originally thought to have been a chapel, archaeological investigations suggest that it was actually a domestic dwelling. With the realignment of the river, this has been left to fall victim to coastal erosion. At this stage, the path turns south to follow the river inland with **views to Cley** ahead.

4 The path reaches a junction where you can either climb the bank opposite and bear left along a raised path or follow a lower path to the right. The raised path offers excellent views of **Cley Windmill**, but the lower parallel path is more suitable for wheelchairs and buggies.

The two paths meet just before a gate. Go through the opening at the side. If you have a wheelchair or buggy, cross the road to join the footpath and head left into Cley. If you don't, turn immediately left to walk along a narrow, raised path with shallow steps down to the road. Having crossed the River Glaven at the outfall sluice, walk past the shops to a road T-junction.

5 Turn right at the T-junction and follow the footpath along the left-hand side of the Holt Road. Look for a narrow alleyway that runs between the houses called Alexanders and West View. Follow this past traditional flint cottages until you reach a back road. Turn right and proceed past Cley Village Hall. Those with wider wheelchairs or buggies may want to avoid the alleyway and continue along the Holt Road until they reach a road to the left, signposted Church Lane, Fairstead and Village Hall. Walk up this road

BLAKENEY SPIT

The spit at Blakeney has been formed by longshore drift. Waves strike the coast at an angle and carry shingle obliquely up the beach. As the waves retreat seawards they drag the material directly back down the beach, ready to be caught up by the next incoming waves. In this way, the shingle gradually moves along the coastline. At the mouth of the River Glaven, it has been taken westwards out to sea and been deposited to form a low-lying shingle ridge. This has created a sheltered inland area where salt marshes have developed.

CLEY CHURCH

The cathedral-like church at Cley is dedicated to St Margaret of Antioch, patron saint of sailors and midwives. The tower is 13th century and the main church 14th century. Originally conceived on a magnificent scale, plans were curtailed in 1348 after the Black Death drastically reduced the local population. Carvings abound, particularly in the porch, and many tombstones feature skulls, bones and maritime symbols.

MARSH SAMPHIRE

Marsh samphire is a delicacy that grows in the muddiest parts of Norfolk's salt marshes. Also known as glasswort, a reference to its former use in glassmaking, it can be picked at low tide, but should be harvested sparingly, as it's a winter food for birds. Wash the samphire and boil for up to 15 minutes. Drain and serve with vinegar or melted butter.

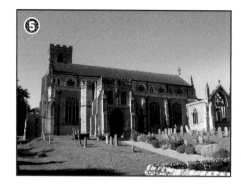

and at the top, with the village hall opposite, turn right onto the back road.

Carry on up a slight incline, past allotments on the left and houses on the right, until the road narrows. In front of Knoll House, take a right fork down a small track leading to a gate into the churchyard. Walk around the tower to the other side of the church. From the main entrance of **St Margaret's Church** walk to the front gate, which looks out over Cley Green.

As its full name suggests, Cley was once very much 'next-the-Sea'. In the Middle Ages, the Glaven was an estuary rather than a river and Cley was a thriving port on its shoreline. The area that is now Cley Green was a busy harbour, visited by ships from all over Europe, bringing spices, cloth and coal to trade for wool, corn and malt. At the beginning of the 17th century, however, the harbour began to silt up. In the shelter of the shingle banks that now form Blakeney spit, salt marshes grew up, restricting the tidal flow and causing the Glaven to narrow and become shallower. This was exacerbated by local landowners who built up banks to enable the surrounding land to be used for grazing without the risk of flooding.

In 1612, when a devastating fire swept through Cley and destroyed 117 dwellings, the inhabitants took the opportunity to rebuild the houses further north and establish a new harbour closer to deep water. Although this prolonged Cley's existence as a port, by the end of the 19th century further

silting, an increase in the size of ships and the development of the rail network saw its eventual demise.

6 With the church and the Three Swallows pub behind you, walk straight across the triangular green to a small road. Walk south along this road, which has views of Wiveton Church over fields to the right. At the crossroads, take a right turn towards Wiveton. The road crosses over the River Glaven at Wiveton Bridge. This was built 700 years ago and is one of the oldest in England still carrying traffic. The road continues up a gentle incline and emerges at the village green, with the church on the right.

Wiveton, like Cley, was once a busy port on the Glaven Estuary, but it, too, succumbed to silting and ceased trading in the 17th century. Evidence of its seafaring past can be found at the parish church of **St Mary the Virgin**. The marks on the churchyard wall were made when sailors, wanting to secure their ships more thoroughly, threw anchors into the church grounds. Inside the church drawings of ships from the 15th and 16th centuries can be found etched into the stone pillars and walls.

7 Carry on past the village green and, with the Wiveton Bell pub to your right, walk along the road signposted to Blakeney. There is a short stretch of path which runs along the right-hand side of the road, but most of this section involves walking along the tarmac. The road gradually ascends, affording **views back to both Wiveton and Cley.**

Continue past the entrance to Bee Hive Farm and on into Blakeney. As you walk past the church of St Nicholas, the view is

dominated by the main tower, which, at over 100 feet, is one of the highest in Norfolk. Also look out for the curious little tower at the other end of the church, where a light would burn to guide ships safely into Blakeney harbour. From the church, carry on up the road until you reach the junction with the main coast road.

8 Opposite you will see two roads heading straight on. Cross over and walk down the left-hand one. This is **Blakeney High Street**, which slopes down to the quayside. Continue down past the Old Customs House on the right and the Old Bakery on the left. Look out for the narrow alleyways, known locally as 'lokes', which lead off the High Street into pretty courtyards flanked by flint cottages. As you turn a corner, the view over Blakeney harbour opens up. Cross the road to return to the car park.

North Norfolk Coast

The river and coast path have been diverted to the south since this OS map was compiled (see text). The yellow route indicates the new riverside path.

ADVICE

The route is accessible to people with wheelchairs or buggies, although parts of the coast path are a little uneven. There are also places where the path becomes quite narrow and there are shallow steps at one point, which can be avoided by a short diversion. The route is generally flat, free from stiles and accessible all year round. Refreshments are available in Blakeney and at Cley-next-the-Sea.

PARKING

There is a pay-and-display car park, as well as toilets, by Blakeney harbour.

START

From Blakeney harbour car park, follow the Norfolk Coast Path signpost.

CONTACT DETAILS

Cromer Tourist Information Centre, Prince of Wales Road, Cromer, Norfolk NR27 9HS
t: 0871 2003071
e: cromertic@north-norfolk.gov.uk
w: nationaltrail.co.uk/peddarsway

Great Yarmouth
Beyond the bucket and spade

EASY

ACCESS

2½ MILES

3:00

THE FIRST NELSON'S COLUMN

In South Denes, off the route of the walk, you will see Nelson's Column. The 144-foot-tall monument was erected in 1819, 24 years before the one in London's Trafalgar Square. Fund-raising for the pillar started in Nelson's lifetime to mark his victory at the Nile and his return to Norfolk. However, by the time the £10,000 had been raised, the seafaring hero had died. Instead of being dedicated to the battle, it became a memorial.

Dig beyond Great Yarmouth's bucket and spade image to unearth an intriguing past on a walk around what was, in the 1400s, England's fifth richest town.

The walk starts at the jetty, just off **Marine Parade**. The sea which laps against the jetty has attracted a huge number of visitors to Great Yarmouth over the years. The flashing lights and noises that emanate from the amusement arcades lining the town's Golden Mile illuminates its position as one of the country's top seaside resorts. But this loud and brash façade should not overshadow the town's compelling history.

In the past, the coast off Great Yarmouth was one of the best places in the world to fish for herring. In Victorian times, Great Yarmouth's standing as a holiday destination

became as important as its herring industry. But now, as you start the walk, look out to sea and try to picture the scene some 200 years ago. The wind is caught in the sails of a battleship making its way back to the town after the Battle of Copenhagen. A victorious Admiral Nelson is leading his crew back to the safety of his home county after destroying the Danish fleet.

2 Turn left along the seafront and on Marine Parade you cannot miss the imposing frontage of the Windmill Theatre. The entertainment of visitors has been taken seriously here ever since this theatre, then known as the Gem, became home to the country's first electric picture house in 1908. It was just one of five theatres and cinemas that lined up on the imposing seafront at the turn of the century, but it was health concerns that originally attracted visitors to the town in the 18th century. 'Taking the waters' became fashionable for those who could afford to head to the coast to drink and soak in the salt water, after a doctor wrote a pamphlet hailing it as a cure for a range of illnesses in 1759.

When the Norwich to Yarmouth railway opened in 1844 the less well off joined in the fun. The beach offered a change of scenery from smoky, land-locked, industrial towns.

3 From the Windmill walk down the seafront, turn right at the Hotel Elizabeth and head along Camperdown. At the top, go right and immediately left up Malakoff Street. Head straight on until you reach the **town wall**.

This is one of the best-preserved town walls in the country. Built in the late 1200s on the orders of Henry II, to repel unwanted invaders, its construction was dogged by lack of money. A hundred years later, it almost became the town's death sentence. People lived in cramped rows of houses inside the perimeter, which led to appalling conditions where an enemy, far more deadly than the French, thrived. The Black Death spread like wildfire through Yarmouth's close-knit community, wiping out two-thirds of its population.

4 From the wall, turn right, heading to St Peter's Road. When you get there take a left, then turn right at the end of the road into King Street.

For nearly a thousand years, if you lived in Great Yarmouth you would almost certainly have lived in the rows of streets crammed inside the town wall. The lanes of houses were unique and drew the eye of famous visitors, including Charles Dickens. The writer lived in Great Yarmouth in 1848, using many of the town's characters and landmarks as inspiration for his classic novel *David Copperfield*.

But for people who lived in the dark, tightly packed lanes life was not much fun. The paths streamed with sewage. The rows were designed to run east to west, so the easterly winds would shift the pong. The pebble pathways were sloped downwards towards the harbour so all the mess would end up in the sea when it rained. The 145 rows were crammed so close together that one alley measured just 18 inches across at its narrowest point. Inevitably, neighbours living cheek to jowl caused problems and there was little privacy.

From 1930 onwards sections of the rows were condemned. Ironically, two days of German air raids in the early months of the Second World War were more effective, destroying some 2000 houses. Today, restored examples of the rows and houses can be seen at English Heritage's **Row 111** and the Old Merchant's House on South Quay.

5 Moving on, turn left into Yarmouth Way. Straight ahead of you is the **Tolhouse**. A stay here, in one of the country's oldest gaols, was a grim business. In the 1400s men, women and children were piled together in two filthy, dark dungeons teeming with mice and rats, where they could fester for up to ten years while waiting for their trial. Punishments such as flogging, branding with hot irons, being locked in public stocks and hanging were regularly meted out.

In 1875 the Tolhouse was closed and the remaining inmates were moved to the new

SHUT THAT DOOR

Most of the 145 rows were between 3 and 5 feet wide so just opening a door could injure innocent passers-by. Eventually a law was passed to make householders reverse the hinges on their doors so they would open inwards instead. Bailiffs were sent as enforcers, and those who ignored the order were fined and their door was nailed shut until they complied.

ANGEL OF MERCY

The gruelling conditions inside the Tolhouse started to improve when a Caister woman, Sarah Martin, took an interest in the prisoners' plight. The religious dressmaker began going to the gaol in 1818 to hold Sunday services to inspire the inmates to improve their lives. She balanced her spiritual advice with practical help, and taught them how to read and write, as well as make items like books and spoons to sell. Martin gave the prisoners a sense of purpose and boosted their pride by using the profits to buy clothes.

THE ROCK FACTORY

Docwra's, which is based in Regent Road on the way to the pier, has been making souvenir candy in Great Yarmouth since 1896 and is believed to be the world's biggest rock factory. At its height in the 1960s, when the Midlands' factories shut down for the holidays, the shop would open seven days a week, from 8.00 a.m. until 11.00 p.m., to cope with visitors wanting souvenirs to take back home.

Norwich Prison. Although badly damaged by a bomb in 1941, the former merchant's house was rebuilt and re-opened as a museum.

6 Walk past the pillar-box and the library. Turn right at the main road and cross at the crossing. Walk until the cobbles become paving and you can see the Star Hotel on the right. You are now on South Quay.

Around 1900 Great Yarmouth was one the world's leading herring ports, taking in some 800,000 fish each day from over 600 boats. South Quay was brim full of **fisher girls**, many from Scotland, gutting and grading fish in one of the industry's most gruelling jobs.

This strenuous life on shore was less perilous though than that of the fishermen, who could be at sea for weeks. Most of the vessels headed for Smith's Knoll, around 30 miles north-east of Great Yarmouth, which was one of the best places in the world to

Copyright © Great Yarmouth Museum

catch the herring, or the 'silver darlings' as they were called.

Gradually the herring industry dwindled due to over-fishing, foreign competition and the collapse of overseas markets, and by the mid 1960s it was virtually gone.

7 From South Quay head for Haven Bridge. Cross the road towards the Star Hotel. Turn right and then left into Regent Street. At the end of Regent Street turn left and walk through the **Market Place** until you can see St Nicholas's Church.

In its 19th-century heyday thousands of people attended services here. One day in 1827 rumours of gruesome events at the church began to spread. Parishioners began to suspect their vicar was indulging in more than God's work. In the dead of night, one of them braved the graveyard to dig up his dead relative to find an empty coffin. The next day the chilling news spread across the market and the families of the recently deceased went to exhume the graves, many of which were found to be empty.

It turned out that the vicar had a nephew at a hospital in London who wanted the cadavers for his anatomy classes. The bodies were snatched from their graves after dark, and smuggled in barrels down the coast and up along the Thames.

8 Go past the Market Gates shopping centre, turn left at BHS, go straight down Regent Road and onto **Britannia Pier**. Today the pier marks the top of the Golden Mile, Great Yarmouth's pleasure parade. But at the start of the 1940s the mood was much more sombre.

Great Yarmouth was bombed more than any other coastal town in the country. It was an easy target – German planes flew in over the North Sea, dropped their payload and got out before they could be intercepted. A shattering 8000 bombs were dropped here.

The war changed the face of Great Yarmouth forever. The Hollywood Cinema with its 40-yard-high tower was sent for scrap, and the Royal Aquarium too, the beach was mined and the pier was blown in two to stop it being used as a landing point for Hitler's forces.

But that wasn't the only trauma that the wooden pier has endured since its opening in 1858. It was struck by a boat and burned down four times in the first 80 years of its life. Given the run of disasters, surprisingly, the pier's ballroom and pavilion stood firm during the war – only to be burned down again in 1954.

When you've had enough of the pier, turn right and walk along Marine Parade, back to where you started.

Great Yarmouth

ADVICE

The walk is mostly along public footpaths with some cobbled areas and a wooden jetty and is accessible for wheelchair users and parents with pushchairs. The circular route has eight stages and you can enjoy the experience through your mobile phone with characters from the past sharing their tales of the town's history.

PARKING

The nearest pay-and-display car park is next to the jetty. There are also car parks along the seafront opposite Zen Nightclub, at the Marina Centre, opposite Pleasure Beach Gardens, near the bowling greens on North Drive and by the Waterways, also on North Drive.

START

The walk starts at the jetty, just off Marine Parade, which forms the town's seafront.

CONTACT DETAILS

Great Yarmouth Tourist Information Centre, Maritime House, 25 Marine Parade, Great Yarmouth, Norfolk NR30 2EN
t: 01493 846345
f: 01493 858588
e: tourism@great-yarmouth.gov.uk
w: www.great-yarmouth.co.uk

Ordnance Survey Explorer Map number OL40
© Crown Copyright 2008

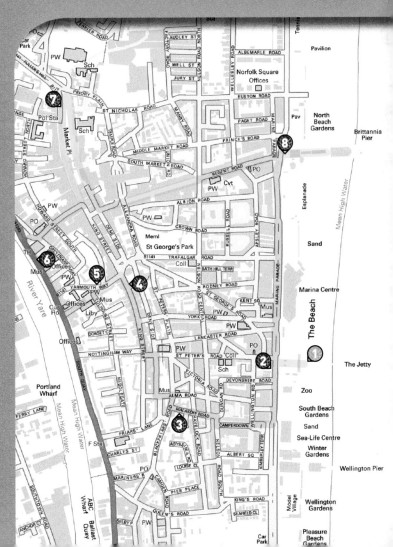

Orford
Peace and War

MEDIUM

ACCESS

6 MILES

6:00

This walk provides opportunities to enjoy the beauty and peace of Orford Ness today and to relive its extraordinary history of military activity.

Orford is a pleasant village located close to the mouth of the River Ore, literally at the end of the B1084. Yet despite the peaceful atmosphere, the threads of military defence and war run through its history. Orford Castle was built and fortified by Henry II in the 12th century as a defence against a rebellious local baron. Orford Ness, a shingle spit about 13 miles long, was an important military site used from the First World War to 1985. Currently Orford Ness is owned by the National Trust, which is restoring the shingle vegetation that was damaged by the military activity.

1 Begin the walk by turning right out of the National Trust car park into Quay Street and walk straight ahead at the crossroads into Church Street. On your right you will see St Bartholomew's Church, which was built in the 14th century. Follow the road around to Market Hill and the market square. From here you will see the magnificent **Orford Castle**, set on a grassy mound. Only the polygonal keep remains, but if time allows it is worth a visit. Now turn left down Castle Hill, into Broad Street, both pretty streets of red brick houses. Turn right back into Quay Street and walk to the Quay.

② The ferry crossing to Orford Ness is very short, but you may spot some interesting birds on the way. From June to early September you should see house martins swooping over the water and catching insects. At low tide there may be waders on the mud.

Once on the Ness, a guide will explain the layout of the site and warn of the **dangers of unexploded ordnance**. There are three footpaths, coded red, blue and green, marked by red, blue and green arrows respectively, so it is difficult to get lost.

③ To walk the entire blue trail, leave the jetty following the arrows on the path, with the river wall on your left. The wall was built in the late 12th century, probably by Henry II, to create grazing land from the salt marsh. Here you find fragments of salt marsh where seawater seeps through the wall. In August and September you can see red-tinted glasswort and annual sea blight. Sea asters are prolific here, too, and very pretty with their blue daisy-like flowers. Turn right at the fork following the blue trail and now you are walking between the old airfield and the 'new' **grazing marsh**. This is an atmospheric wetland habitat with clumps of sea club rush in shallow pools, frequented by wading birds such as oystercatchers, lapwings, curlew, and avocets. The glistening mud is marked by the footprints of wading birds, ducks and geese. If you examine the reeds closely, you may find some of the rare spiders that inhabit the Ness.

④ Continue following the blue path, turning left at the fork where you find a cluster of **historic buildings**. The building in front of you was part of the officers' mess in the First World War, and later used as the receiver building by Robert Watson-Watt's radar research team during the Second World War. Close by on the right-hand side of the path are the remains of a building used as sick quarters in the First World War, now flooded by brackish water. This shallow pool is a favourite spot for migrating common and green sandpipers in spring and autumn. On the left of the path is a long, roofless Nissen hut, the remains of a model atomic bombing range built in 1954. A little further on, on the right-hand side of the path, you can see a long, dark building at right angles to the path, which was the transmitter building for Watt's radar research team.

Walk on and you will soon see the Information Building on your right. It is worth going in and reading about the fascinating history of the site and viewing the archive photographs. Then, continuing the walk, turn right past the Information Building and on your right you can see a large brick and cement building with a corrugated iron roof.

ORFORD NESS LIGHTHOUSE

Orford Ness lighthouse was built in 1792 by Lord Braybroke and purchased by Trinity House in 1837. Trained lighthouse keepers were employed and they lived in cottages at the base of the tower. Another lighthouse, built in 1720, closer to the sea, was used as a low light and both were important for guiding ships through the treacherous sand and shingle banks. In 1887 the low light was engulfed by the sea so more lights were incorporated into the remaining lighthouse. During both world wars the lighthouse was administered by the Naval Authority, which operated it only when needed. The lighthouse has used electricity since 1959 and was automated in 1965.

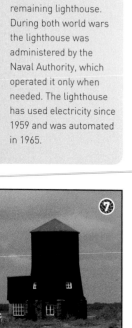

This was used as a barracks in the First World War and subsequently converted into a store.

5 By now you can see a grassy river wall in front of you. Climb the steps to the viewing platform to see **Stony Ditch**, a tidal creek flowing into the River Ore. At low tide the exposed mud is a feeding ground for black-headed and herring gulls, and waders, curlew, little egrets, oystercatchers and avocets. Return to the path, passing an accommodation block used by the National Trust. On your left there is the only public convenience on the Ness. Walk on until you see the Bailey bridge on your right, which crosses Stony Ditch. From the bridge you can see the pagodas, test sites for atomic bombs in the 1950s.

Cross the bridge and turn left. Now you are walking on a shingle path and you should see plants such as yellow-horned poppy, sea campion, thrift and sea pea. The 'horns' of the yellow-horned poppy grow up to a foot long and split when ripe, freeing large numbers of tiny seeds. The sea pea is rare, but easy to spot when the pink flowers are out from June to August. It is important to walk only on the paths because of unexploded ordnance and to protect the shingle plants.

By now you can see the lighthouse in the distance and the shingle seems to stretch for miles. The building on your left was completed in 1933 and designed to hold the instruments for recording the flight paths of bombs dropped from aircraft, part of the research effort for investigating and improving the aerodynamics of atomic bombs, though they were never fitted with warheads. Climb the steps to the top of the building for a good view of Orford Ness and the depressions in the shingle left after bombs were dropped onto it.

6 Head towards the **lighthouse**, following the shingle path. Next to the lighthouse is a scruffy building, which was a coastguard watch house built in the mid-19th century. The watch house was abandoned in the early 1960s and has deteriorated much since then. Now the path moves to the shingle beach itself, which is more uncomfortable for walking on, but the closeness of the North Sea and the waves crashing on the shingle provide a refreshing change. Follow the red arrows and soon you see a red arrow pointing inland, where the path is a bit easier for walking. Walk past the police tower where you can see mats of sea campion growing in the shingle.

7 The shingle path takes you to a building, completed in 1928 and used for testing a 'rotating loop' **navigation beacon**. Close by on the right is the beacon's powerhouse. Drifts of red valerian thrive here, which, when in flower, provide a bright contrast to the muted beiges of the shingle. Now turn left, following the asphalt track towards the Impact Facility. This vast construction of concrete and shingle was used in the 1960s to test the time-delay fuses fitted in atomic bombs. The fuse was placed on a rocket-propelled sledge on rail tracks and the sledge was shot into a concrete wall to mimic an atomic bomb hitting the ground.

8 A little further on is Laboratory 1, built in 1956 for drop testing, in which bombs were subjected to extremely high accelerations and collisions. You can also view the outer casing of an atomic bomb. Outside Laboratory 1, the end of the red trail, there is a good view of the **pagodas** and other laboratories, which have reinforced concrete roofs designed to absorb vertical blasts.

Retrace your steps following the red trail to the black beacon and turn left following the trail to the Bailey bridge. There are more opportunities here to view shingle plants and in the far distance the aerials on the Cobra Mist site can be seen, which are the transmitters for the BBC World Service. Turn left after the Bailey bridge, following the red trail back and turning right on the red trail for views of the old grazing marsh on your right. The red trail takes you back to the jetty, hopefully in time to catch the ferry back to Orford.

Orford

ADVICE

Wear a pair of good walking shoes as the shingle is uncomfortable to walk on. Orford Ness is very exposed with little shade, so in summer, sun hats, sun cream and a supply of water are essential. In spring and autumn the Ness can be very cold with high winds, so warm waterproof clothing is essential. If you want to view the birds and mammals, binoculars are very useful. If you plan to spend much of the day on the Ness, you need to take food and drink with you as there are no facilities there apart from toilets. It is especially important to check the timetable for the ferry to Orford Ness, as although sailings are frequent from July to the end of September, they decline to Saturdays only in October and from April to June. Plan your walk so that you are back at the jetty by 5.00 p.m. at the latest, as this is the final crossing for the day. It is advisable to buy the guidebook too, as this provides a map and important information.

PARKING

There is a National Trust pay-and-display car park in Orford.

START

The walk starts from the National Trust car park in Orford.

CONTACT DETAILS

Orford Ness National Nature Reserve (National Trust), Orford Quay, Orford, Woodbridge, Suffolk IP12 2NU
t: **01394 450900**
f: **01394 450900**
e: **orfordness@nationaltrust.org.uk**
w: **nationaltrust.org.uk**

Ordnance Survey Explorer Map number 212
© Crown Copyright 2008

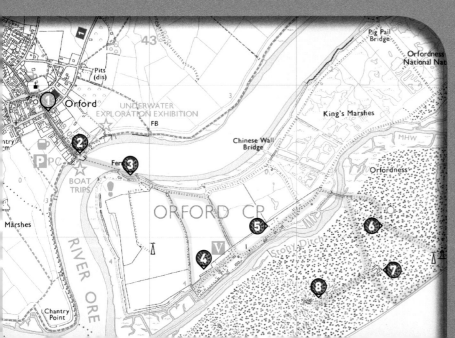

Walton-on-the-Naze
One of southern England's hidden gems

MEDIUM

ACCESS

3 MILES

2:30

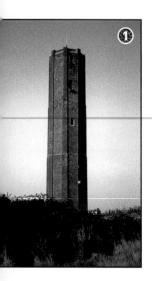

①

The Naze is an area of headland rich with ancient fossils and unique wildlife. Once a playground for Victorian holidaymakers, this constantly changing landscape has always been involved in a bitter battle for survival against the sea.

①

This circular walk starts and finishes at the historic **Naze tower**, taking in some of the sights and sounds of nearby Walton Town and traversing some of the most beautiful unspoilt countryside in Essex.

Start the walk on the cliff-top in the lee of the impressive tower, a monument that has dominated the headland for the last 250 years. Built in 1721 by Trinity House, its original purpose was to act as a marker for ships approaching Harwich harbour, a duty it still performs today. Known locally as the Landmark, the land around it is being lost to

the sea at the alarming rate of about six feet a year and the tower has become a symbol of survival. Recently purchased by a local resident, it is now open to the public and affords breathtaking views of the surrounding area.

Make your way northwards along the grassy cliff-tops, heeding the warning notices to keep well clear of the unstable edge.

The Naze – from the word 'naes' meaning nose (the shape of the land as it juts out to sea) – was originally farmland, then a privately owned golf course. During the Second World War the area was requisitioned as part of the coastal defences. In 1967 it was bought by the local council and has been an area of public open space ever since, enjoyed by locals and thousands of visitors every year. The dense brambles and hawthorn provide a haven for hundreds of species of animals and insects, and an important landing area for migrating birds.

2 The 70-foot high **fossil-rich cliffs** along the top of which you are walking began forming over 54 million years ago when the land lay beneath a warm sea. Rivers flowed into it bringing mud and silt, which eventually became compacted and formed what we now know to be London clay and which makes up the base of the Naze. The clay is overlaid with Red Craig, a sandy deposit a mere two million years old.

One of the finest geographical sites in Britain, the cliffs are popular with fossil hunters, who come to search for traces of the marine life that would have swum in the sea millions of years ago. The cliffs here have also produced some of the best bird fossils ever seen.

It is easier to take a closer look at the cliffs from the beach at the end of the walk (see **8**) where there is plenty of evidence of the large-scale cliff erosion. Two pillboxes up on the Naze, which served as lookout posts during the Second World War, have since fallen onto the beach below. As the path descends across the top of the lower cliffs, you can see Harwich in the distance ahead of you.

3 Follow the path through the trees at the far end of the Naze until you meet a tarmac path. Bear left along the top of the embankment. On your left is the **John Weston nature reserve**. Named after the leading Essex naturalist who was warden of the reserve until his death in 1984, this area of blackthorn, bramble thickets and rough grassland is home to hundreds of birds and animals.

A walk through the reserve might be rewarded with views of nesting birds, including lapwing, redshank, and sedge and reed warblers. Beyond the reserve is the 1½-mile-long shingle beach, ending at Stone Point, which is an important nesting site for little terns and other shorebirds, and an important landfall for migrants. It also attracts a good variety of winter visitors.

4 Continue on the tarmac path until it ends, and bear left again to continue along a grassy path – still on the top of an embankment – above Cormorant Creek and the surrounding marshland. The Naze tower dominates the skyline to the left. Ahead of you lie the **Walton Backwaters** and the clatter of the masts from the many yachts moored there will greet you as you come up from the nature reserve.

HOLDING BACK THE SEA

With what seems to be the judgement of Solomon, the local council has seen fit to strengthen the southern part of the beach with groynes and a seawall, while the northern part has been left undefended. This seems to reflect local opinion, which is split between shoring up coastal defences and leaving nature to do as it pleases.

The Backwaters are a series of tidal creeks, mudflats, islands, salt marshes and marsh grounds. The public footpath that runs along much of the seawall is a great vantage point for the many species of wildlife, including birds, insects and even seals.

⑤ Continue heading south along the path in the direction of Walton Town keeping the broader expanses of the Walton Backwaters on the right. You will soon come to the remains of a once grand **iron foundry**.

In the early 19th century a family called Warner moved to Walton from London. Father John, who owned and ran an iron foundry in London, saw potential for the development of a seaside resort and began work immediately. When his grandson Robert took over the business in 1869 he purchased an extra piece of land and built a second foundry.

The foundry was the main source of employment in the town in the late 19th and early 20th centuries, with between 150 and 300 workers. They made everything from iron seats and portable cooking boilers to wind-powered pumps. They also produced material for the Indian Railways and were bell-founders for Queen Victoria, King Edward VII and King George V. The factory gradually declined, though it continued production until it closed in the 1960s.

From the foundry take the grassy public footpath on your left (as opposed to the gravel driveway). You will come out of the footpath on Naze Park Road.

⑥ The original village of Walton, once known as Eadolfenaesse and also Waltonia, is now 9 miles out to sea on the west rocks, its church having finally fallen into the sea in 1798.

The **new town**'s development began in earnest in the early 19th century when the new-found economic confidence of the Victorians prompted a rise in the popularity of seaside holidays. The arrival of the railway and the building of the pier further encouraged visitors, who arrived in their thousands from nearby cities like Ipswich and London. The building of hotels and other amenities gave rise to what is described as Walton's heyday when it even rivalled Southend for popularity. Today the town retains a strong flavour of that golden era, despite a recent regeneration project.

Walk up Naze Park Road towards the sea, go across the park, walk along the seafront adjacent to Cliff Parade and then along the Greensward. For a shorter route carry on up Naze Park Road into Old Hall Lane and turn off into Sunny Point, which will lead you back to the Naze.

⑦ Walton-on-the-Naze has been home to a busy **coastguard and lifeboat station** since the 1880s. Today's lifeboat is moored near the end of the pier and is the only one in Britain to have a permanent mooring in the open sea. When needed, the crew cycle the length of the pier and then use a small launch to access the lifeboat.

The old lifeboat house is now the Walton Maritime Museum, which has a collection of local memorabilia, including displays on natural sciences, weapons and war, personalities, science and technology, social history and the maritime history of the town.

When you leave the museum walk across the road and down onto **Jubilee beach**. Walk northwards along a beach lined with beach huts and cafes that also boasts the second longest pier in Britain (Southend being the longest). Miles of sandy beaches still prove as popular as ever during the summer months.

When you come to a cafe walk around it to the right and carry on along the beach or the seawall towards the Naze tower, which you can get up to via some stairs. If you can't manage stairs go up the slope to the left of the cafe and return to the Naze along the cliff-top.

Before returning to the starting point you should take the opportunity to search the cliffs for fossils.

WARNING

Do not attempt to get to the fossil cliffs at high tide as there is a risk of being cut off. Take heed of local notices and check the tide news at bbc.co.uk/weather/coast/tides/

Walton-on-the-Naze

ADVICE

A large part of the walk is on grass, but is accessible for wheelchairs or buggies. Part of the walk also takes in the cliff-edge where extreme care should be taken.

PARKING

There is a pay-and-display car park at the Naze, or free car parking in nearby streets, if available.

START

You start the walk in the shadow of the Naze tower.

CONTACT DETAILS

Walton Community Project,
61 High Street,
Walton-on-the-Naze,
Essex CO14 8AG
t: 01255 677006
f: 01452 311899
e: info@walton-on-the-naze.com
w: walton-on-the-naze.com

Ordnance Survey Explorer Map number 184
© Crown Copyright 2008

Picture credits and acknowledgements

**BBC BOOKS WOULD LIKE TO THANK THE FOLLOWING
FOR PROVIDING TEXT AND PHOTOGRAPHS:**

Dr Patricia Ash (Dover, Orford); Dr Eric Bowers (Lynton and
Lynmouth, Bristol Docks, Glamorgan Heritage Coast);
Dr Mark Brandon (South Pembrokeshire, North Pembrokeshire);
Dr Mike Dodd (Isle of Skye, Rubha Reidh, Faraid Head, Durness);
Mrs Glynda Easterbrook (The Jurassic Coast, Lyme Regis,
Flamborough Head); Dr Ian Johnston (Dublin, Achill Island, The
Causeway Coast, Portpatrick, Isle of Arran, Mull); Dr Dick Morris
(Spey Bay and Moray, Aberdeen, Dundee); Dr David Robinson
(Saxon Shore Way, Brighton, Isle of Wight); Dr David Sharp,
Alice Fraser, Julie Fraser and Tamsyn Fraser (King's Lynn,
North Norfolk Coast); Dr Sheila Stubbles (Devon's South Hams,
South-East Cornwall, North Cornwall); Dr Janet Sumner
(Llyn Peninsula, Anglesey, The Great Orme); Christopher Tinker
(Tresco); Dr Paul Williams (Northumberland, Durham).

The following walks have been adapted from the BBC *Coast*
website (www.bbc.co.uk/coast) and the text and pictures
are © copyright BBC (unless stated opposite*): Rotherhithe,
Portsmouth, Falmouth, Gloucester's Waterways, Cardiff Bay,
Maritime Merseyside, Whitehaven, Belfast, Culzean, Glasgow,
Stromness, Perth and Tayside, Old Leith, Borders Coastal Walk,
Whitby, Hull, Great Yarmouth, Walton-on-the-Naze.

* Thanks are also due to the following who provided additional
photographs as detailed: pages 24–7 (all photos) Julian Flanders,
except page 24 (bottom) © Mary Rose Trust; page 50 (both)
© National Maritime Museum Cornwall; page 64 (opening
picture) Carole McDonald; pages 68 (bottom) and 69 (right)
© Gloucester City Museum; page 70 (right) *Gloucestershire Echo*;
pages 72–5 (all photos) Carole McDonald; page 109 (top left
and bottom) © Aidan O'Rourke; page 110 (left) © Getty Images;
page 111 (right) © Aidan O'Rourke; pages 112–15 (all photos)
Mark Robertson; pages 116–19 (all pictures) Aidan O'Rourke;
pages 128–31 © National Trust for Scotland (main picture
and vinery by David Robinson, pictures 1 and 4 by Harvey Wood,
2 by E. Lamb, 3 and 6 by Isla Robertson, 5 by Glyn Satterley,
7 by Kathy Collins, 8 by Gordon Riddle); page 156 (bottom)
© Patricia Long; page 176 © Carmen Mardiros; page 177
(all pictures) © Reproduced with acknowledgement to
Peter Stubbs (www.edinphoto.org.uk); page 178 (top, right and
bottom) © Carmen Mardiros; page 178 (left) © Peter Stubbs;
page 193 (bottom left), 194 (top left) and 195 © Whitby Museum;
pages 200–3 © Hull City Council; page 214 (bottom) Great
Yarmouth Museums, pages 212–15 © Zippix.

While every effort has been made to trace and acknowledge all
copyright holders, BBC Books would like to apologize should there
have been any errors or omissions.